Lecture Notes in Mathematics Vol. 493
ISBN 978-3-540-07420-5

Franz W. Kamber and Philippe M. Tondeur

Foliated Bundles and Characteristic Classes

Erratum

Formula (5.100) on page 128 should read as follows:

$$C \;:\; \dim \widehat{P} = \operatorname{rank} \mathfrak{g} - \operatorname{rank} \mathfrak{h} \qquad \text{(Cartan pair)}\,,$$

Lecture Notes in Mathematics

Edited by A. Dold and B. Eckmann

493

Franz W. Kamber
Philippe Tondeur

Foliated Bundles and Characteristic Classes

Springer-Verlag
Berlin · Heidelberg · New York 1975

Authors
Prof. Franz W. Kamber
Prof. Philippe Tondeur
Department of Mathematics
University of Illinois
at Urbana-Champaign
Urbana, Illinois 61801
USA

AMS Subject Classifications (1970): 57D20, 57D30, 55F99

ISBN 3-540-07420-1 Springer-Verlag Berlin · Heidelberg · New York
ISBN 0-387-07420-1 Springer-Verlag New York · Heidelberg · Berlin

Printed in Germany
Offsetdruck: Julius Beltz, Hemsbach/Bergstr.

ACKNOWLEDGMENTS

These notes are the revised text of lectures held in summer 1973 at the University of Heidelberg during a sabbatical leave from the University of Illinois, and again in fall 1973 and spring 1975 at the University of Illinois. Our work on these topics was supported by several grants from the National Science Foundation. We would like to take this opportunity to thank B. Eckmann for the hospitality extended to us at the Forschungs-Institut für Mathematik of the Eidg. Techn. Hochschule, where we worked and lectured on this subject during the summers of 1971, 1972 and spring 1973.

Thanks are due to V. Vold for a critical reading of the manuscript and to J. Largent for her careful typing.

April 1975 Franz W. Kamber Philippe Tondeur

LIST OF CHAPTERS

CONTENTS

INTRODUCTION

The authors have introduced in the last few years a construction of characteristic classes for foliated bundles which provides among other things a construction of characteristic invariants of foliations. The purpose of these lectures is to present this construction, and to interprete and compute these new invariants in various geometric contexts.

The basic concept in this theory is a foliated bundle. This is a principal bundle P with a foliation on the base space M, and a partial connection on P which is only defined along the leaves of the foliation on M, and which has zero curvature (definition 2.1). For the trivial foliation of M by points this concept reduces to an ordinary principal bundle P and then our construction of characteristic classes reduces to the ordinary Chern-Weil construction. Our construction was inspired by the work of Chern-Simons [CS 1]. For the trivial foliation of M consisting of one single leaf the concept of a foliated bundle reduces to a flat bundle. Our work on flat bundles (summarized in [KT 1]) was one of our motivations for the introduction of the concept of a foliated bundle in [KT 2,3]. This concept allows the simultaneous discussion of ordinary bundles, bundles with an infinitesimal or global group action, flat bundles and normal bundles of foliations. For a foliation, the canonical foliated bundle structure on the frames normal to the foliation gives rise to characteristic invariants attached to the foliation. For this situation, the construction in [KT 4-7] is one of the various independently discovered approaches to characteristic invariants by Bernstein-Rosenfeld [BR 1] [BR 2], Bott-Haefliger [B 3] [H 5] [BH], Godbillon-Vey [GV], Malgrange (not published) and the authors.

In the first two chapters of these lectures we discuss the basic geometric concepts and illustrate these with many examples. The list of contents is so as to require no detailed comments. We wish just to make the following conceptual remark. The consideration of the tangent bundle of manifolds leads with necessity to the consideration of arbitrary vectorbundles. In the same way the consideration of the normal bundle of foliations with its flat partial Bott-connection (definition 2.15) leads to the consideration of arbitrary foliated vectorbundles (and foliated principal bundles). This is in particular indispensible for functorial considerations and for the other examples of foliated bundles mentioned. This point of view is indeed just a generalization of the former viewpoint, since for the trivial foliation of the base space M by points the concept of a foliated bundle reduces to the ordinary concept of a bundle.

In chapter 3 we discuss the construction of characteristic classes for the special case of a flat bundle, in which case it is particularly simple and of independent interest. The basic idea is that a flat connection ω in a principal G-bundle $P \longrightarrow M$ defines a homomorphism of differential algebras (proposition 3.3)

$$\Lambda^{\cdot}\underline{\underline{g}}^{*} \longrightarrow \Omega^{\cdot}(P).$$

The domain is the Chevalley-Eilenberg complex of the Lie algebra $\underline{\underline{g}}$ of G (definition 3.2). The range is the DeRham complex of differential forms on P. The map is the multiplicative extension of the canonical map $\underline{\underline{g}}^{*} \longrightarrow \Omega^{1}(P)$ defined by dualizing ω. The induced cohomology map (corollary 3.10)

$$k_{*} : H^{\cdot}(\underline{\underline{g}}) \longrightarrow H^{\cdot}_{DR}(P)$$

furnishes cohomology classes on P. To get cohomology classes in the base space M, we need the further data of a closed subgroup $H \subset G$ and an H-reduction P' of the G-bundle P given by a section s of $P/H \longrightarrow M$. The characteristic homomorphism for these data is then the composition (theorem 3.30)

$$\Delta_* = s^* \circ k_* : H^{\bullet}(\underline{\underline{g}},H) \longrightarrow H^{\bullet}_{DR}(M).$$

The relative Lie algebra cohomology $H^{\bullet}(\underline{\underline{g}},H)$ is the cohomology of the complex of H-basic elements $(\Lambda^{\bullet}\underline{\underline{g}}^*)_H$ in $\Lambda^{\bullet}\underline{\underline{g}}^*$. This terminology as well as the concept of $\underline{\underline{g}}$-DG-and G-DG-algebras are explained carefully in chapter 3. The list of contents will help to locate the definitions in the text.

The first type of non-trivial invariants as just described is given at the end of chapter 4 starting with 4.79. Chapter 6 is devoted in its entirety to examples of characteristic classes for flat bundles in various contexts.

The construction of the generalized characteristic homomorphism for a foliated bundle in chapter 4 is a natural generalization of the construction just explained for flat bundles. In the presence of curvature the map $\Lambda^{\bullet}\underline{\underline{g}}^* \longrightarrow \Omega^{\bullet}(P)$ defined by a connection ω in P is no longer a homomorphism of differential algebras. The map to consider then is the Weil homomorphism (proposition 4.14)

$$k(\omega) : W^{\bullet}(\underline{\underline{g}}) \longrightarrow \Omega^{\bullet}(P)$$

canonically extending the map on the Chevalley-Eilenberg complex. The Weil algebra $W^{\bullet}(\underline{\underline{g}})$ (definition 4.2, lemma 4.3) has been introduced in [CA]. It is canonically filtered (definition 4.21), and $\Omega^{\bullet}(P)$ carries a filtration defined by the foliation on the base space M (definition 4.22). For any connection ω on P which is adapted to the foliated bundle structure (definition in 1.35) the Weil homomorphism $k(\omega)$ is a filtration preserving homomorphism of differential algebras (theorem 4.23). This leads in particular to a vanishing theorem (4.27) for $k(\omega)$ on a filtration ideal of $W^{\cdot}(\underline{\underline{g}})$ characterized by the codimension q of the foliation on M. This generalizes Bott's vanishing theorem [B 1] [B 2] on the ordinary characteristic classes of the normal bundle of a foliation. It also leads to a homomorphism defined on the truncated Weil algebra (corollary 4.39). To construct invariants in the base space M of the foliated bundle P, one proceeds as before. For a closed subgroup $H \subset G$ and an H-reduction P' of P the generalized characteristic homomorphism is then a map

$$\Delta_* : H^{\cdot}(W(\underline{\underline{g}},H)_q) \longrightarrow H^{\cdot}_{DR}(M)$$

whose properties are stated in theorem 4.43.

This construction in the case of the normal bundle of a foliation is compared with the constructions of Godbillon-Vey and Bott-Haefliger in 4.48. The geometric significance of the new invariants as obstruction classes is discussed in 4.51.

The authors wish to point out that in this construction the generalized characteristic homomorphism is just the induced map in total cohomology of the two spectral sequences associated to the relative truncated Weil algebra $W^{\cdot}(\underline{\underline{g}},H)_q$ and the DeRham complex $\Omega^{\cdot}(M)$. The characteristic homomorphism of a foliated bundle appears therefore more generally as a comparison map between two spectral sequences. The induced homomorphisms on the E_r-terms are what we call the derived characteristic homomorphisms. They are briefly discussed in 4.50 together with the spectral sequence defined by the filtration of $\Omega^{\cdot}(M)$ via the foliation on M.

The computation of the cohomology algebras $H^{\cdot}(W(\underline{\underline{g}},H)_q)$ is the subject of chapter 5. The basic construction is a complex realizing this cohomology (Theorem 5.85). The introduction of a semi-simplicial model for the Weil algebra is an essential tool for these computations. It plays also a fundamental role for the construction of homotopies between the characteristic homomorphisms defined for different adapted connections on the cochain level. The introduction of semi-simplicial models for the Weil algebra plays further an essential role if one wants to apply the methods developed so far in the context of complex analytic manifolds and algebraic varieties. This is the subject of chapter 8. The outline in 8.1 explains the natural procedure to follow.

Chapter 7 is devoted to examples of this generalized construction of characteristic classes. Particularly striking results are obtained for foliations of groups by subgroups. We refer to the text and [KT 10] for details. The variety of geometric contexts for the examples of chapters 6 and 7 should be the best argument for the usefulness of the concepts and constructions presented in these lectures.

TABLE OF CONTENTS

1. FOLIATIONS

In this chapter we discuss a few examples of smooth foliations of smooth manifolds. A foliation of codimension q on a manifold M is a partition $\{\mathcal{L}_\alpha\}_{\alpha\in A}$ of M into connected leaves $\mathcal{L}_\alpha$ with the following property. For every point in M there is a neighborhood U and a chart $f = (f_1,\dots,f_n) : U \longrightarrow \mathbb{R}^n$, $n = \dim M$ such that for each leaf $\mathcal{L}_\alpha$ the components of $\mathcal{L}_\alpha \cap U$ are described by the equations $f_{n-q+1} = \text{constant}, \dots, f_n = \text{constant}$. Thus locally the leaves of a codimension q foliation look like a set of parallel planes of codimension q in Euclidean space.

Foliations appear as families of solutions of systems of differential equations. The study of foliations is the study of the global behavior of such solutions. E.g. a first order differential equation is a vectorfield. For a vectorfield without zeros the local solutions, i.e. the orbits of the flow generated by the vectorfield, form a 1-dimensional foliation. The study of the global aspects of the orbits of a vectorfield goes back to Poincaré. The concept of a foliation has been introduced by Ehresmann and Reeb [ER] [R]. For an introduction and survey of the subject of foliations we refer to the article by Lawson [L]. For the narrower selection of topics here discussed we refer to Bott's lectures [B3].

For the purposes of these lectures we are primarily interested in the infinitesimal object associated to a foliation, i.e. the subbundle of the tangent bundle T_M of M consisting of all vectors tangent to the leaves of the foliation. In the example of a non-zero vectorfield mentioned before, this means that we are concentrating our attention on the vectorfield itself rather than the solutions of the vectorfield. To describe the critical property satisfied by the bundle of vectors tangent to a foliation, we need the following concept. Let E be a smooth subbundle of T_M. Then $\Gamma(U,E)$ denotes the smooth sections of E over an open set $U \subset M$.

1.1 DEFINITION. _The bundle_ $E \subset T_M$ _is involutive, if for any open_ $U \subset M$

$$X,Y \in \Gamma(U,E) \Longrightarrow [X,Y] \in \Gamma(U,E).$$

It is clear that the bundle of vectors tangent to a foliation is involutive. By the Theorem of Frobenius (Thm. 1.18 below) an involutive subbundle conversely is the bundle of vectors tangent to a well-defined foliation. We use the letter L for an involutive subbundle $L \subset T_M$, and use in these lectures the term foliation even when we wish only to talk about the infinitesimal object L.

The quotient

$$Q = T_M/L \tag{1.2}$$

is a well-defined vectorbundle of dimension

$$q = \dim Q = \text{codimension of } L. \tag{1.3}$$

This is the transversal or normal bundle of L. We have the exact sequence

$$0 \longrightarrow L \longrightarrow T_M \longrightarrow Q \longrightarrow 0 \tag{1.4}$$

of vectorbundles on M. The equivalent exact sequence of dual vectorbundles reads

$$0 \longleftarrow L^* \longleftarrow T_M^* \longleftarrow Q^* \longleftarrow 0, \tag{1.5}$$

where the dual E^* of a vectorbundle E is defined as the bundle of homomorphisms into the trivial line bundle.

1.6 EXAMPLE. The simplest example is given by a non-zero vectorfield X on M. Let $L \subset T_M$ be the line bundle with fiber L_x spanned by X_x for $x \in M$. It is involutive. In fact any line bundle $L \subset T_M$ is involutive, as follows from the formula

$$(1.7) \qquad [X,gY] = g[X,Y] + Xg\,.\,Y$$

valid for any local tangent vectorfields X,Y and smooth function g on M. For a line bundle L and local trivializing section Z we have namely by (1.7) for $X = fZ$, $Y = gZ$

$$\begin{aligned}[X,Y] &= g[X,Z] + (Xg).Z \\ &= -g(f[Z,Z] + Zf.Z) + Xg.Z = (-g.Zf + Xg).Z\end{aligned}$$

which is again a section of L.

1.8 EXAMPLE. A nowhere zero 1-form ω on M defines a line bundle $Q^* \subset T^*_M$ with fiber Q^*_x spanned by ω_x for $x \in M$. The bundle L defined via (1.4) (1.5) is given by

$$(1.9) \qquad L_x = \ker \omega_x \subset T_xM$$

L is involutive if and only if $\omega[X,Y] = 0$ for local vectorfields X,Y such that $\omega(X) = 0$, $\omega(Y) = 0$. In view of the formula $d\omega(X,Y) = X\omega(Y) - Y\omega(X) - \omega[X,Y]$ this is equivalent to

$$(1.10) \quad d\omega(X,Y) = 0 \quad \text{for } X,Y \text{ such that } \omega(X) = 0,\ \omega(Y) = 0\,.$$

If $L \subset T_M$ is more generally any involutive subbundle of codimension 1, it is locally given by a 1-form ω as described above. Note that (1.10) is equivalent to the local representability of $d\omega$ as

$$d\omega = \omega \wedge \alpha$$

for a local 1-form α. To see this we chose a local framing of T_M^* by 1-forms $\alpha_1,\dots,\alpha_n$ with $\alpha_1 = \omega$. Then locally the 2-form $d\omega$ has a representation

$$d\omega = \sum_{i<j} d\omega(X_i,X_j)\alpha_i \wedge \alpha_j$$

where the X_i form a dual local framing of T_M, i.e. $\alpha_i(X_j) = \delta_{ij}$. But the only non-zero terms occur by (1.10) for $i = 1$, so that as claimed

$$d\omega = \sum_{j>1} d\omega(X_1,X_j)\omega \wedge \alpha_j = \omega \wedge \alpha$$

with $\alpha = \sum_{j>1} d\omega(X_1,X_j)\alpha_j$. It is a remarkable result of Thurston [TH 3] that on a compact manifold M a codimension 1 foliation exists if and only if the Euler number $\chi(M)$ equals zero.

Let $E \subset T_M$ be any subbundle of codimension q. Locally E can again be described by

$$E_x = \ker \omega_x \subset T_x$$

where ω is a local 1-form on M with values in a q-dimensional vectorspace V. Choosing a basis in V, ω can be described by q ordinary 1-forms, and

$$E_x = \bigcap_{i=1}^{q} \ker \omega_i(x) \subset T_x, \tag{1.11}$$

where the 1-forms $\omega_1,\dots,\omega_q$ are defined on an open neighborhood

U of x and form a basis for the local sections of Q^* over U.

1.12 DEFINITION. The bundle $E \subset T_M$ of codimension q is integrable, if locally there exists real smooth functions $f_1,\ldots,f_q$ such that

(1.13) $$E_x = \bigcap_{i=1}^{q} \ker df_i(x) \subset T_x.$$

The functions $f_1,\ldots,f_q$ define on an open set $U \subset M$ a smooth map $f = (f_1,\ldots,f_q) : U \longrightarrow \mathbb{R}^q$. Since the derivative $df(x) : T_xM \longrightarrow \mathbb{R}^q$ has by assumption precisely E_x as kernel, it is surjective and $f : U \longrightarrow \mathbb{R}^q$ is a submersion. Equivalently the differentials $df_1(x),\ldots,df_q(x)$ are linearly independent. The inverse images under f partition U into (not necessarily connected) submanifolds, and E_x appears as tangent space through the unique submanifold through x. Thus U is foliated by the connected components of these submanifolds. These local foliations of neighborhoods for different U match, and thus an integrable subbundle $E \subset T_M$ defines a foliation of M.

1.14 EXAMPLE. The simplest examples of integrable bundles are obtained for submersions $f : M \longrightarrow N$, i.e. smooth maps such that $df_x : T_xM \longrightarrow T_{f(x)}N$ is surjective. Then

(1.15) $$T(f)_x = \ker df_x$$

defines an integrable subbundle $T(f) \subset T_M$, the tangent bundle along the "fibers" of f. Note that in this case the normal bundle Q is the pull-back of the tangent bundle of N

(1.16) $$Q = f^* T_N$$

and the exact sequence (1.4) in this case is as follows:

(1.17) $$0 \longrightarrow T(f) \longrightarrow T_M \longrightarrow f^*T_N \longrightarrow 0.$$

Definition (1.12) requires an integrable subbundle to be locally of this form. Note that the space of leaves of the foliation defined by $T(f)$ is a space over N and coincides with N precisely when all fibers of f are connected.

Let ω be a nowhere zero 1-form on M which is closed. The codimension one subbundle $L \subset T_M$ defined by the vectors annihilated by the form ω is by (1.10) involutive. It is an interesting theorem of Tischler [TS] that for compact M the existence of such a 1-form implies the existence of a fibration $f : M \longrightarrow S^1$ such that the foliations defined by L and $T(f)$ are close.

The relation between the infinitesimal condition of involutivity (1.1) and the integrability condition (1.12) is given by the following classical result.

1.18 THEOREM OF FROBENIUS. *Let $E \subset T_M$ be a smooth subbundle of codimension q, described locally as in (1.11) by linearly independent 1-forms $\omega_1, \ldots, \omega_q$. Then the following conditions are equivalent:*

(i) *E is involutive;*

(ii) *there exist local 1-forms α_{ij} $(i,j = 1, \ldots, q)$ such that*

$$d\omega_i = \sum_{j=1}^{q} \alpha_{ij} \wedge \omega_j, \quad i = 1, \ldots, q;$$

(iii) *the ideal defined by $\omega_1, \ldots, \omega_q$ in the (local) De Rham complex is a differential ideal;*

(iv) *E is integrable;*

(v) *there exist local functions $f_1, \ldots, f_q$ and g_{ij} $(i,j = 1, \ldots q)$ such that*

$$\omega_i = \sum_{j=1}^{q} g_{ij}\, df_j \quad (i = 1, \ldots, q).$$

In (ii) and (v) the stated equalities hold in a neighborhood of every point x from the original open set on which $\omega_1, \ldots, \omega_q$ are given. Condition (ii) states that the ideal defined by $\omega_1, \ldots, \omega_q$ in the De Rham complex of all differential forms (on a possibly smaller open set) contains all differentials $d\omega_i$. This clearly implies (iii), which states that this ideal is closed under d. In (v) the matrix of functions (g_{ij}) is necessarily invertible.

Any of the conditions in Frobenius' Theorem characterizes a foliation. The partition into leaves is obtained by taking maximal integral manifolds of the subbundle $E \subset T_M$. These are more precisely foliations without singularities.

1.19. It is often desirable to consider foliations having singularities, as e.g. defined by a vectorfield having a non-empty set of zeros, or a smooth map $f : M \longrightarrow N$ which fails to be a submersion at a certain set of points in M.

1.20 EXAMPLE. Consider the vectorfield

$$X = -x_2 \frac{\partial}{\partial x_1} + x_1 \frac{\partial}{\partial x_2}$$

on $\mathbb{R}^2$. The origin is a zero of X. The integral curve of X through $x = (x_1, x_2)$ is the circle

$$\gamma(t) = (r \cos t, r \sin t), \; r = (x_1^2 + x_2^2)^{1/2}$$

The origin is a singular integral curve $(r = 0)$.

Consider the differential form

$$\omega = x_1 \, dx_1 + x_2 \, dx_2$$

on $\mathbb{R}^2$. Note that $\omega = df$, where $f(x) = \frac{1}{2}(x_1^2 + x_2^2)$. The origin is a zero of ω, i.e. a critical point of $f : \mathbb{R}^2 \longrightarrow \mathbb{R}$. This is a singular leaf in the foliation defined by f.

On $\mathbb{R}^2 - \{0\}$ the vectorfield X and the 1-form ω are mutual annihilators and the foliation can equivalently be described by the line bundle $L \subset T_{\mathbb{R}^2 - \{0\}}$ spanned by X and the line bundle $Q^* \subset T^*_{\mathbb{R}^2 - \{0\}}$ spanned by ω. On $\mathbb{R}^2$ this is not so anymore.

1.21 REMARKS ON FOLIATIONS WITH SINGULARITIES. The character of the singularities of foliations given by sets of vectorfields or 1-forms is not necessarily the same. One needs to distinguish the two cases. The natural approach is to use the following sheaf terminology [GT]. Let $\underline{O}_M$ denote the sheaf of smooth functions on M. Then the sections of the tangent bundle T_M form an $\underline{O}_M$-module $\underline{T}_M$ and more generally the sections of any subbundle $E \subset T_M$ an $\underline{O}_M$-submodule $\underline{E} \subset \underline{T}_M$. The converse is not true, unless $\underline{T}_M/\underline{E}$ is a locally free $\underline{O}_M$-module, in which case we can reconstruct a subbundle $E \subset T_M$ whose sections form $\underline{E}$. Similar facts hold for subbundles of the dual tangent bundle T^*_M and $\underline{O}_M$-submodules of its sheaf of sections $\underline{T}^*_M$, i.e. the sheaf of 1-forms on M. One has then the following natural definitions.

1.22 DEFINITION. An $\underline{O}_M$-submodule $\underline{E} \subset \underline{T}_M$ is involutive, if (1.1) holds.

1.23 DEFINITION. An $\underline{O}_M$-submodule $\underline{Q}^* \subset \underline{T}^*_M$ is involutive, if for every locally generating set of 1-forms of $\underline{Q}^*$ the ideal generated in the DeRham complex is a differential ideal.

If $\underline{Q}^* \subset \underline{T}^*_M$ is in particular locally generated by a finite number $\omega_1,\ldots,\omega_q$ of 1-forms, this condition is equivalent to condition (ii) in Theorem (1.18).

An $\underline{O}_M$-submodule $\underline{E} \subset \underline{T}_M$ defines an $\underline{O}_M$-submodule $\underline{Q}^* \subset \underline{T}^*_M$ as the annihilator sheaf

$$\underline{Q}^* = \underline{\mathrm{Ann}}(\underline{L}), \tag{1.24}$$

i.e. the subsheaf of local 1-forms on M vanishing on $\underline{L}$. Conversely $\underline{Q}^* \subset \underline{T}^*_M$ defines an $\underline{O}_M$-submodule $\underline{L} \subset \underline{T}_M$ as the annihilator sheaf

$$\underline{L} = \underline{\mathrm{Ann}}(\underline{Q}^*), \tag{1.25}$$

i.e. the subsheaf of local vectorfields on M annihilating all forms in $\underline{Q}^*$. But the relation need not be reflexive. It is so in the non-singular case.

The natural integrability condition for submodules $\underline{E} \subset \underline{T}_M$ is a relaxation of condition (1.13), where the local functions f_i may fail to define a submersion on a certain singular set. The natural integrability condition for submodules $\underline{Q}^* \subset \underline{T}^*{}_M$ is a relaxation of condition (v) in (1.18), where the matrix (g_{ij}) may fail to be invertible on a certain singular set. There exist several generalizations of Frobenius theorem, comparing involutivity and integrability conditions either for $\underline{O}_M$-submodules of

$\underline{T}_M$ or of $\underline{T}^*_M$ (generally speaking integrability implies involutivity but not conversely).

For the purposes of these lectures, the infinitesimal involutivity conditions are the only one needed for both types of submodules. The integrability conditions never enter directly in our arguments. Since we discuss characteristic invariants of foliations, it means that these are necessary conditions for the corresponding infinitesimal objects already under the simple assumption of involutivity (not necessarily integrability). For the sake of simplicity we formulate the whole subject in terms of non-singular foliations. But many results apply equally well for singular foliations.

The terminology discussed above is as explained useful for the inclusion of certain types of singular foliations even in the smooth context. But this terminclogy is indispensable to be able to carry out the theory as here presented in the complex analytic context or for algebraic varieties (see [KT 6,7]).

Another reason for the introduction of singular foliations is Haefliger's homotopy-theoretic approach to foliations (see [H 1] and [H 3] , [H 4]). If one wishes to classify isomorphism classes of foliations on a space X by homotopy classes of maps from X into a universal (classifying) space, the definition of a foliation has to be loosened. The starting point is the following observation. A codimension q foliation on a manifold M is given by an open covering $\mathfrak{U} = \{U_i\}_{i \in I}$ and submersions $f_i : U_i \longrightarrow \mathbb{R}^q$ for $i \in I$, satisfying the following properties. For each $i,j \in I$ and each $x \in U_i \cap U_j$ there is a diffeomorphism γ^x_{ij} from a neighborhood of $f_i(x)$ to a neighborhood of $f_j(x)$ such that

$f_j = \gamma^x_{ji} f_i$ on a neighborhood of x. For each $x \in U_i \cap U_j \cap U_k$ $\gamma^x_{ki} = \gamma^x_{kj} \circ \gamma^x_{ji}$. These cocycle conditions guarantee that the local foliations on the U_i defined by the submersions f_i piece together to a global foliation. The transition functions of the normal bundle are obtained by the differentials $g^x_{ij} = d\gamma^x_{ij}$.

Haefliger defines now singular foliations on any space X by allowing the local maps f_i to be any continuous maps. The 1-cocycle γ is for each $i,j \in I$ a continuous map $\gamma_{ij} : U_i \cap U_j \longrightarrow \Gamma_q$ into the germs of local diffeomorphism of $\mathbb{R}^q$ with its sheaf topology. A Haefliger-structure on a space is an equivalence class of such cocycles under refinement of the covering $\mathfrak{U}$ in the usual Čech-sense. Haefliger-structures then pull back under continuous maps $f : X \longrightarrow Y$. Two Haefliger-structures on a space X are concordant if there exists a Haefliger-structure on a cylinder $X \times [0,1]$ inducing on top and bottom the two given Haefliger-structures. A beautiful result of Haefliger is then that concordance classes of Haefliger-structures on (paracompact) spaces X are classified precisely by homotopy classes of continuous maps $X \longrightarrow B\Gamma_q$ into a classifying space $B\Gamma_q$. In fact Haefliger fits all this into a perfectly elegant general theory which generalizes the classification theory of fiber bundles with structural group.

We discuss further examples of foliations.

1.26 EXAMPLE. Let $G \times M \longrightarrow M$ be a smooth action of a Lie group G on the manifold M. The orbits of G define a foliation of M, if the action is almost free, i.e. every isotropy group is discrete. In general the orbits, which correspond to conjugacy classes of closed subgroups, are not necessarily of the same dimension. The orbits with maximal dimension (principal orbits) can be considered regular leaves and the lower dimensional orbits singular leaves of a (singular) foliation. A simple example is the circle action

$$S^1 \times \mathbb{C} \longrightarrow \mathbb{C}$$

$$(t,z) \longmapsto e^{it}z$$

where $S^1 = \mathbb{R}/2\pi\mathbb{Z}$. The origin is a unique singular orbit.

1.27 EXAMPLE. A non-singular foliation is in particular obtained for an action which is simply transitive on the orbits. An example is the foliation on the total space of a principal bundle by the orbits of the structural group. We will consider in particular the foliation of a Lie group by the cosets modulo a subgroup.

1.28 HOMOGENEOUS FOLIATIONS. Let more generally $H \subset G \subset \overline{G}$ be subgroups of a Lie group $\overline{G}$ with H closed in $\overline{G}$ (hence closed in G). Then the foliation of $\overline{G}$ by the left cosets of G induces a foliation of the homogeneous space $\overline{G}/H$. We call such a foliation homogeneous. The canonical left action of $\overline{G}$ on $\overline{G}/H$ induces automorphisms of the homogeneous foliation.

1.29 LOCALLY HOMOGENEOUS FOLIATIONS. Let $\Gamma \subset \overline{G}$ be in addition to the data in the previous example a discrete subgroup operating properly discontinuously and without fixed points on $\overline{G}/H$, so that the double coset space $\Gamma\backslash\overline{G}/H$ is a manifold. Then the $\overline{G}$-invariant foliation induced by G on $\overline{G}/H$ induces a foliation on $\Gamma\backslash\overline{G}/H$. We call such foliations locally homogeneous.

1.30 FLAT CONNECTIONS. Let $P \longrightarrow M$ be a principal bundle with structure group a Lie group G. A $\underline{\underline{g}}$-valued connection form ω on P defines the subbundle $H \subset T_P$ of horizontal spaces on P by

$$H_u = \ker \omega_u \subset T_u P.$$

Recall the structure equation

$$\Omega = d\omega + \frac{1}{2}[\omega,\omega]$$

where by definition for vectorfields X,Y on P

$$[\omega,\omega](X,Y) = 2[\omega(X),\omega(Y)].$$

(Note that this normalization differs by a factor 2 from the one in [KN 1], as does the normalization of the exterior derivative $d\omega(X,Y) = X.\omega(Y) - Y.\omega(X) - \omega[X,Y]$.) The structure equation shows that for horizontal X,Y

$$\Omega(X,Y) = 0 \iff d\omega(X,Y) = 0$$

The argument leading to equation (1.10) in example (1.8) shows that H is involutive if and only if this condition is satisfied, i.e. if and only if the curvature $\Omega = 0$. Thus the horizontal spaces of a connection define a foliation if and only if the connection is flat.

It may be of interest to observe that in this case the local submersions defining the foliation are locally defined maps $f : P \longrightarrow G$ such that the canonical $\underline{\underline{g}}$-valued Maurer-Cartan form θ on G (given by $\theta(x) = x$ for $x \in \underline{\underline{g}}$) satisfies $f^*\theta = \omega$. In fact the Maurer-Cartan equation $d\theta + \frac{1}{2}[\theta,\theta] = 0$ on G implies $d\omega + \frac{1}{2}[\omega,\omega] = 0$ for $f : P \longrightarrow G$ and $\omega = f^*\theta$, and it is well-known that the latter equation is in turn sufficient for the existence of local maps $f : P \longrightarrow G$ such that $f^*\theta = \omega$.

1.31 PARTIAL CONNECTIONS. Let $\overline{G} \longrightarrow \overline{P} \longrightarrow M$ be a principal bundle with connection $\overline{\omega}$. The subbundle $\overline{H} \subset T_{\overline{P}}$ of horizontal spaces is invariant under the right action of $\overline{G}$ on $\overline{P}$. It follows that for any closed subgroup $G \subset \overline{G}$ the subbundle $\overline{H}$ induces on the quotient $P = \overline{P}/G$ a subbundle $H \subset T_P$. (For the case $G = \overline{G}$ this is the tangent bundle T_M, where $M = \overline{P}/\overline{G}$.) The projection $\overline{P} \longrightarrow P$ is the projection of a G-principal bundle. But $\overline{\omega}$ does not define a connection in this bundle. The subbundle $\overline{H} \subset T_{\overline{P}}$ maps into a subbundle $H \subset T_P$ which need not equal T_P. Thus $\overline{H}$ in the G-bundle $\overline{P} \longrightarrow P$ need no longer be a full complement to the fiber G. It is a partial connection in the sense of the following definition.

1.32 DEFINITION. <u>A partial connection in a principal bundle</u> $G \longrightarrow P \longrightarrow M$ <u>is a subbundle</u> $H \subset T_P$ <u>such that</u>

(i) $H_u \cap G_u = \{0\}$ <u>for every</u> u <u>and</u> G_u <u>the tangentspace to the fiber through</u> u,

(ii) $H_{ug} = (R_g)_* H_u$ <u>for every</u> $u \in P$ <u>and</u> $g \in G$, <u>where</u> R_g <u>denotes the right action of</u> g <u>on</u> P.

The G-equivariance of $H \subset T_P$ guarantees that there is a well-defined subbundle $E \subset T_M$ onto which H projects under $\pi : P \longrightarrow M$, i.e. $\pi_* H_u = E_x$ for every $u \in P$ such that $\pi(u) = x$. The notion of a horizontal vectorfield is generalized in the obvious way: a partially horizontal vectorfield is a vectorfield X on P such that $X_u \in H_u$ for all $u \in P$. For vectorfields X on M belonging to E there is then as usual a unique partially horizontal lift $\tilde{X}$ on P such that $\pi_* \tilde{X}_u = X_{\pi(u)}$.

1.33 ADAPTED CONNECTIONS. Instead of developing a new calculus for partial connections, we found it more convenient to introduce the following concept, which allows then the use of the standard calculus.

1.34 DEFINITION. A connection in the principal bundle $P \longrightarrow M$ is adapted to a given partial connection in P, if the horizontal subspace of the connection contains the subspace given by the partial connection, for each $u \in P$.

Note the following. If ω is an adapted connection, then we need only be given the subbundle $E \subset T_M$. The subbundle $H \subset T_P$ namely is then already defined by

$$H_u = \ker \omega_u \cap \pi_*^{-1} E_{\pi(u)}$$

for $\pi_* : (T_P)_u \longrightarrow (T_M)_{\pi(u)}$.

For the following discussion we recall that the projection $\pi : P \longrightarrow M$ pulls back forms on M to forms on P. In this way the DeRham complex $\Omega^\cdot(M)$ of forms on M generates a subcomplex of the DeRham complex $\Omega^\cdot(P)$ of forms on P, i.e. $\pi^*\Omega^\cdot(M) \subset \Omega^\cdot(P)$. The same holds for the corresponding sheaves of local forms. We use the notation $\Omega_M^\cdot$ for the complex of sheaves of local forms on M and similarly $\Omega_P^\cdot$ on P. With the usual notation $\Gamma(U,-)$ for the sections of a sheaf over an open subset $U \subset M$ we have in particular $\Omega^\cdot(M) = \Gamma(M,\Omega_M^\cdot)$ and $\Omega^\cdot(P) = \Gamma(P,\Omega_P^\cdot)$. The inclusion $\pi^*\Omega_M^\cdot \subset \Omega_P^\cdot$ is then the sheaf version of the inclusion mentioned before. In particular the sheaf $\underline{Q}^* \subset \Omega_M^1$ of local 1-forms annihilating the vectorfields belonging to a subbundle $E \subset T_M$, appears via π^* as $\pi^*\underline{Q} \subset \Omega_P^1$.

Let now ω and ω' be two connections which are adapted to a given partial connection on P. The difference $\varphi = \omega' - \omega$ is a tensorial 1-form on P of type Ad, i.e. φ satisfies the conditions [KN 1, p. 75]

$$R_g^* \varphi = Ad(g^{-1})\varphi \quad \text{for } g \in G$$

$$\varphi(X) = 0 \quad \text{for } X \text{ a vertical vectorfield on } P.$$

Let $\underline{Q}^*$ be the sheaf of local 1-forms on M, annihilating the vectorfields belonging to the subbundle $E \subset T_M$, which is the projection of $H \subset T_P$ to M. It is clear that $\pi^*\underline{Q}^* \subset \Omega_P^1$ consists of the local 1-forms on P vanishing on vertical and partially horizontal vectorfields. It follows then that for every $\alpha \in \underline{\underline{g}}^*$ the 1-form $\alpha\varphi$ is a global section of $\pi^*\underline{Q}^*$, i.e. $\alpha\varphi \in \Gamma(P, \pi^*\underline{Q}^*)$. Conversely a tensorial 1-form φ on P of type Ad with this property for all $\alpha \in \underline{\underline{g}}^*$ added to an adapted connection ω will furnish another adapted connection $\omega' = \omega + \varphi$.

The upshot of this discussion is the following result [KT 7].

1.35 PROPOSITION. *A partial connection in the principal G-bundle $P \longrightarrow M$ is characterized by the following data.*

(i) A subbundle $E \subset T_M$;

(ii) a class of connections $\{\omega_i\}_{i\in I}$ on P (called adapted connections) such that the differences $\varphi_{ij} = \omega_j - \omega_i$ are tensorial 1-forms of type Ad with the property: for each $\alpha \in \underline{\underline{g}}^*$ the form $\alpha\varphi_{ij} \in \Gamma(P, \pi^*\underline{Q}^*)$, where Q^* is the sheaf of local 1-forms on M annihilating the vectorfields belonging to E.

The point of this remark is that given $E \subset T_M$, the partial connection in P is given completely by a single connection (called adapted). All other adapted connections are already determined by a single one.

1.36 FLAT PARTIAL CONNECTIONS. Now we turn to the discussion of the flatness of a partial connection.

1.37 DEFINITION. <u>A partial connection</u> $H \subset T_P$ <u>in a principal bundle</u> P <u>is flat, if the subbundle</u> $H \subset T_P$ <u>is involutive</u>.

Note that then the subbundle $L \subset T_M$ obtained from the G-equivariant subbundle H by the projection $P \longrightarrow M$ is also involutive. Therefore a flat partial connection in P is firstly a foliation $L \subset T_M$ of the base space, and secondly a lift of L to a G-equivariant foliation H of P. Note that for two vectorfields X,Y on M belonging to L and their partially horizontal lifts $\tilde{X},\tilde{Y}$ on P belonging to H we have then

$$\widetilde{[X,Y]} = [\tilde{X},\tilde{Y}]. \tag{1.38}$$

This condition is obviously equivalent to the flatness of the partial connection.

1.39 PROPOSITION. <u>Let</u> $H \subset T_P$ <u>be a partial connection in</u> P, <u>such that the subbundle</u> $L \subset T_M$ <u>obtained by projection on</u> M <u>is involutive. The following conditions are equivalent</u>:

(i) H <u>is flat</u>;

(ii) <u>for all adapted connections</u> ω <u>in</u> P <u>the curvature</u> Ω <u>satisfies</u>

$$\Omega(X,Y) = 0 \quad \text{for all partially horizontal } X,Y; \tag{1.40}$$

(iii) <u>for one adapted connection</u> ω <u>in</u> P <u>the curvature</u> Ω <u>satisfies</u> (1.40).

Proof. (i) $\Longleftrightarrow$ (ii): For the curvature Ω of any adapted connection ω and partially horizontal X,Y we have

(1.41) $$\Omega(X,Y) = d\omega(X,Y) = -\omega[X,Y].$$

Thus $[X,Y] \in H$ implies $\Omega(X,Y) = 0$. Assume conversely that H is not involutive, i.e. for some $u \in P$ and some partially horizontal X,Y the bracket $[X,Y]_u \notin H_u$. Choose an adapted connection ω such that its horizontal space at u does not contain $[X,Y]_u$. Then $\Omega_u(X,Y) \neq 0$ by (1.41) and (ii) fails.

(ii) $\Longleftrightarrow$ (iii): Trivially (ii) $\Longrightarrow$ (iii). To prove the converse, assume (1.40) holds for ω and let $\omega' = \omega + \varphi$ be any other adapted connection. Then the curvature Ω' of ω' is of the form

$$\Omega' = \Omega + [\omega,\varphi] + \frac{1}{2}[\varphi,\varphi] + d\varphi$$

By the characterization of φ in (1.35), (ii) it follows that (1.40) holds also for ω. □

We further wish to find an expression for the flatness of a partial connection which does not involve partially horizontal vectorfields on P, but solely the foliation $L \subset T_M$ and an adapted connection on P. For this purpose we consider again the annihilator sheaf $\underline{Q}^* \subset \underline{T}^*_M$ on L and its lift $\pi^*\underline{Q}^* \subset \Omega^{\cdot}_P$. Consider the ideal $\pi^*\underline{Q}^*.\Omega^{\cdot}_P$ generated in $\Omega^{\cdot}_P$ by $\pi^*\underline{Q}^*$. In degree 2 it consists of germs of local forms on P which are representable as finite sums

(1.42) $$\sum_i \pi^* \varphi_i' \wedge \varphi_i''$$

where $\varphi_i' \in \underline{Q}^*$ and $\varphi_i'' \in \Omega^1_P$.

1.43 PROPOSITION [KT 6,7]. *Let $P \longrightarrow M$ be equipped with a partial connection, defined by a foliation $L \subset T_M$ and an adapted connection ω in P with curvature Ω. The partial connection is flat if and only if the following properties holds. For every $\alpha \in \underline{\underline{g}}^*$ the 2-form $\alpha\Omega$ on P is locally of the form* (1.42), *i.e.*

$\alpha\Omega \in \Gamma(P,\pi^*\underline{Q}^* \cdot \Omega_P^{\cdot})$.

Proof. We first observe that Ω is a horizontal form, i.e. for every fundamental vectorfield X^* on P defined by an element $x \in \underline{g}$ we have

$$(1.44) \qquad i(X^*)\Omega = \Omega(X^*,-) = 0.$$

This implies that Ω need only be evaluated on horizontal vectorfields. Let $n = \dim M$ and let T_M be locally framed by vectorfields $X_1,\ldots,X_{n-q}$ spanning L and vectorfields $Y_1,\ldots,Y_q$ spanning Q. The partially horizontal vectorfields $\tilde{X}_1,\ldots,\tilde{X}_{n-q}$ and the horizontal vectorfields $\tilde{Y}_1,\ldots,\tilde{Y}_q$ with respect to ω form together with a basis of fundamental vectorfields locally a framing of T_P. Let $\alpha_1,\ldots,\alpha_{n-q}$ and $\beta_1,\ldots,\beta_q$ be the 1-forms belonging to the corresponding dual local framing of T_P^*, so that $\alpha_i(\tilde{X}_j) = \delta_{ij}$, $\beta_i(\tilde{Y}_j) = \delta_{ij}$ and $\alpha_i(\tilde{Y}_j) = 0$, $\beta_i(\tilde{X}_j) = 0$. In particular $\beta_1,\ldots,\beta_q$ are a local framing of π^*Q^*.

Let $\alpha \in \underline{\underline{g}}^*$. In view of (1.44) $\alpha\Omega \in \Omega^2(P)$ has the following local representation:

$$(1.45) \qquad \alpha\Omega = \sum_{i<j} \alpha\Omega(\tilde{X}_i,\tilde{X}_j)\alpha_i \wedge \alpha_j$$

$$+\sum_{i<j} \alpha\Omega(\tilde{Y}_i,\tilde{Y}_j)\beta_i \wedge \beta_j + \sum_{i,j} \alpha\Omega(\tilde{X}_i,\tilde{Y}_j)\alpha_i \wedge \beta_j.$$

The last two sums are in $(\pi^*\underline{Q} \cdot \Omega_P^{\cdot})^2$. It remains to show that the first sum vanishes if and only if the partial connection is flat. But the $\tilde{X}_i$ are partially horizontal vectorfields, so that this is precisely the content of proposition 1.39. □

2. FOLIATED BUNDLES

Let $P \longrightarrow M$ be a principal bundle with structure group G.

2.1 DEFINITION. P is foliated if P is equipped with a flat partial connection.

The partial connection is described either by an involutive subbundle $H \subset T_P$ satisfying the conditions (i) (ii) in definition (1.32), or equivalently by a foliation $L \subset T_M$ together with an adapted connection ω. The flatness of the partial connection is then expressed by the condition in proposition 1.43 for the curvature of ω.

This concept was introduced in [KT 2,3] and used in all our work on the subject. Molino has independently introduced this concept in [MO 1] and uses it extensively in his work.

It makes no conceptual difficulty, and it is for certain purposes desirable to generalize this concept to arbitrary fiber bundles. The requirement of the G-invariance of the foliation on the total space has then to be dropped. In an effort to remain in a framework where the standard calculus of connections can be applied, we refrain from such a generalization in these lectures. The variety of geometric situations discussed in these notes suggest by themselves the broader applicability of our constructions of characteristic invariants, which is the main theme of these lectures.

We point out that the term foliated bundle is used by Haefliger [H 5] for any bundle equipped with a foliation transverse to the fiber and such that the tangent space of the foliation is a full complement to the tangent space of the fiber. In case the partial connection in definition 2.1 has this property, i.e. is a full complement to the fiber, then $P \longrightarrow M$ is a flat bundle in the sense of the next example.

2.2 FLAT BUNDLES. A flat G-bundle $P \longrightarrow M$ is a bundle with a flat connection in P. In this case the foliation of M is the trivial foliation consisting of one single leaf.

We recall the notion of the holonomy of a flat bundle (with respect to a basepoint $u_0 \in P$ over $x_0 \in M$) (see the notes [KT 1] for more details on flat bundles). The holonomy map is a homomorphism $h : \Gamma \longrightarrow G$ of the fundamental group $\Gamma = \pi_1(M)$ into the structural group G of P. For a loop γ at $x_0 \in M$ let $\tilde{\gamma}$ be the unique horizontal lift of γ starting at u_0. The element $h(\gamma)$ is then the unique element in G such that $u_0.h(\gamma)$ is the endpoint of $\tilde{\gamma}$.

The homomorphism $h : \Gamma \longrightarrow G$ allows to completely reconstruct the flat bundle $P \longrightarrow M$ as follows (see [KT 1], proposition 3.1 on p. 10). Let $\tilde{M} \longrightarrow M$ denote the universal covering of M. Γ acts on $\tilde{M}$ by covering transformations. Then Γ acts on $\tilde{M} \times G$ by letting it act on the left factor by covering transformations and on G via h. The orbit space $\tilde{M} \times_\Gamma G$ inherits a right G-action and there is a canonical bijection

$$\tilde{M} \times_\Gamma G \cong P$$

which is G-equivariant and hence a G-bundle isomorphism.

2.3 ORDINARY BUNDLES WITH CONNECTIONS. Let ω be a connection in $P \longrightarrow M$. ω is always adapted to the trivial flat partial connection $H \subset T_P$ given by the 0-bundle. In this case the foliation of M is the trivial foliation by the points of M.

2.4 BUNDLE WITH GROUP ACTION. Let $P \longrightarrow M$ be a G-bundle. Let K be a Lie group with a left action on M lifting to an action on P commuting with the G-action. If the action of K on M is almost free, so is the action on P. The orbit foliation of P turns then P into a foliated bundle.

This situation has an obvious analogon for the infinitesimal action of a Lie algebra on M lifting to an infinitesimal action on P. A simple example is a non-zero vectorfield on M whose flow of automorphisms of M lifts to a flow of bundle maps on P. The following is a typical example.

2.5 G-STRUCTURE WITH GROUP ACTION. Let $G \longrightarrow P \longrightarrow M$ be a G-structure on M, i.e. there exists a homomorphism of Lie groups $G \longrightarrow GL(n)$, $n = \dim M$ and an isomorphism of GL(n)-bundles

$$P \times_G GL(n) \cong F(M)$$

of the extension of P to a GL(n)-bundle with the frame bundle of M.

Let K act almost freely on M by automorphisms of the G-structure. The action of K lifts then to P by bundle maps as in the preceding example.

A simple example of the obvious infinitesimal analogon is a non-zero vectorfield whose flow is a flow of automorphismsof the G-structure. If the vectorfield has zeros on M, the foliations on M and P have singularities. An example is a Killing vectorfield on a Riemannian manifold whose flow preserves the orthogonal frame bundle of M.

2.6 FOLIATED VECTORBUNDLES. A vectorbundle associated to a foliated principal vectorbundle inherits a foliated bundle structure, as we now explain.

Let $P \longrightarrow M$ be a foliated principal G-bundle and $\rho : G \longrightarrow GL(V)$ a representation of G in the vectorspace V. Let $E \longrightarrow M$ be the associated vectorbundle with fiber $V : E = P \times_G V$.

A connection ω in P defines a covariant derivation ∇ in E in the usual way [KN 1, p. 114].

2.7 LEMMA. Let ω and ω' be connections adapted to a partial connection in the bundle P and ∇, ∇' the corresponding covariant derivations in an associated vectorbundle E. For vectorfields X belonging to the subbundle $L \subset T_M$ defined by the partial connection and sections s of E we have $\nabla_X s = \nabla'_X s$.

Proof. Recall that a section s of E can be identified with a function $f : P \longrightarrow V$ such that $f(ug) = \rho(g^{-1})\, f(u)$ for $u \in P$ and $g \in G$, if we set $uf(u) = s(\pi(u))$ for $\pi : P \longrightarrow M$ [KN 1, p. 76]. Here $u \in P$ is viewed as a linear isomorphism $V \longrightarrow E_{\pi(u)}$ onto the fiber of E at $\pi(u)$.

Recall further that for a vectorfield X on M and its horizontal lift $\tilde{X}$ to P with respect to a connection we have the formula [KN 1, p. 115].

$$(2.8) \qquad (\nabla_X s)(x) = u(\tilde{X}\, f(u)) \quad \text{for} \quad x = \pi(u).$$

The lemma now follows from the fact that for a vectorfield X belonging to the subbundle $L \subset T_M$ the partially horizontal lift $\tilde{X}$ is independent of the choice of an adapted connection. □

REMARK. Note that the formula for $\tilde{X}f$ used in this proof makes of course only sense for horizontal vectorfields $\tilde{X}$. For the vertical vectorfield X^* defined by an element $x \in \underline{\underline{g}}$, i.e.

$$X^*_u = \frac{d}{dt}\Big|_{t=0} (u.\exp tx) \qquad \text{for} \quad u \in P$$

we get by differentiation of

$$\rho(\exp(-tx))f(u) = f(u.\exp tx)$$

the formula

(2.9) $$-d\rho(x)\, f(u) = X_u^* f$$

where $d\rho : \underline{\underline{g}} \longrightarrow \underline{\underline{g}}\ell(V)$ is the differential of $\rho : G \longrightarrow GL(V)$.

The notion of a partial connection in a vectorbundle $E \longrightarrow M$ with respect to a foliation L on the base space M is therefore defined directly by a covariant derivation operator $\nabla_X s$ for vectorfields X belonging to L and sections s of E satisfying the usual conditions

(2.10)
$$\nabla_{X+Y}(s) = \nabla_X s + \nabla_Y s$$

$$\nabla_{fX}(x) = f\,\nabla_X s$$

$$\nabla_X(s+t) = \nabla_X s + \nabla_X t$$

$$\nabla_X(fs) = f\,\nabla_X s + Xf.s$$

for vectorfields X,Y belonging to L, a smooth function f and sections s,t of E. The flatness of the partial connection in E is characterized by the formula

(2.11) $$R(X,Y) = \nabla_X \nabla_Y - \nabla_Y \nabla_X - \nabla_{[X,Y]} = 0$$

for vectorfields X,Y belonging to L.

An adapted connection in a foliated principal bundle $P \longrightarrow M$ to which the vectorbundle $E \longrightarrow M$ is associated gives rise to a connection in E extending the partial flat connection in E. A direct definition of an adapted connection in a vector-bundle equipped with a partial connection with respect to a foliation of M is as follows.

2.12 DEFINITION. <u>A connection in the vectorbundle</u> $E \longrightarrow M$ <u>is adapted to a given partial connection with respect to a foliation of</u> M <u>if it extends the given partial connection.</u>

To construct such a connection ∇ on E for a given partial connection ∇' on E, one can proceed as follows. Let g be a Riemannian metric on M. Then the exact sequence $0 \longrightarrow L \longrightarrow T_M \longrightarrow Q \longrightarrow 0$ splits and $T_M \cong L \oplus Q$. If ∇'' denotes any connection in E, define

$$\nabla_X s = \nabla'_{X_L} s + \nabla''_{X_Q} s \tag{2.13}$$

for a section s of E and the canonical decomposition $X = X_L + X_Q$ of a tangent vectorfield X. Then ∇ is immediately verified to be a connection in E. ∇ is clearly adapted to ∇'.

Foliated vectorbundles were introduced by the authors in [KT 2,3] under the name $(\underline{L},\underline{O})$-modules. The name foliated bundles has been used in later papers. The concept of a foliated bundle has independently been introduced and extensively used by Molino in his work [MO 1-5], as previously mentioned.

2.14 NORMAL BUNDLES OF FOLIATIONS. Let $L \subset T_M$ be an involutive subbundle defining a foliation with normal bundle $Q = T_M/L$. The bundle Q is canonically foliated by the partial connection defined by

$$\nabla_X s = p[X,Y] \tag{2.15}$$

for a local section s of Q and local vectorfield X belonging to L. Here Y is a local vectorfield such that $p(Y) = s$ under the projection $p : T_M \longrightarrow Q$. If Y' is another vectorfield such that $p(Y') = s$, then $p(Y - Y') = 0$ and hence $Y - Y'$ is a section of L. But then $p[X,Y] - p[X,Y'] = p[X,Y - Y'] = 0$ and $\nabla_X s$ is indeed well-defined. The holonomy of this connection has been introduced by Ehresmann (see [R]). Bott recognized the effect of ∇ on the characteristic classes of Q [B 1] [B 2] (see 4.36 below). (2.15) is called the Bott connection of Q.

To verify the flatness of the Bott connection, we observe that for vectorfields X,Y belonging to L and a vectorfield Z with $p(Z) = s$, we have by (2.10)

$$\nabla_X \nabla_Y s - \nabla_Y \nabla_X s - \nabla_{[X,Y]} s$$

$$= p[X,[Y,Z]] - p[Y,[X,Z]] - p[[X,Y],Z]$$

$$= p([X,[Y,Z]] + [Y,[Z,X]] + [Z,[X,Y]]) = 0.$$

Any (partial) connection ∇ on Q induces a (partial) connection ∇^* on the dual bundle Q^* by the formula

(2.16) $$(\nabla^*_X \omega)(s) = X\omega(s) - \omega(\nabla_X s)$$

for a vectorfield X (belonging to L), 1-form ω belonging to Q^* and section s of Q. Let again Y be a vectorfield such that $p(Y) = s$ for $p : T_M \longrightarrow Q$. Then for the Bott connection (2.15) the dual connection reads by (2.16)

(2.17) $$(\nabla^*_X \omega)(Y) = X\omega(Y) - \omega[X,Y]$$

since ω vanishes on L. But the canonical Lie derivation of X on 1-forms ω is given by

$$(\theta(X)\omega)(Y) = X\omega(Y) - \omega[X,Y]$$

so that in fact for the Bott connection on Q^*

$$\nabla^*_X \omega = \theta(X)\, \omega .$$

Now we use the identity $\theta(X) = i(X)d + di(X)$, where $i(X)$ denotes the interior product defined by

$$(i(X)\varphi)(X_1, \ldots, X_{q-1}) = \varphi(X, X_1, \ldots, X_{q-1})$$

for a q-form φ and vectorfields $X_1, \ldots, X_{q-1}$ on M. Since ω vanishes on L, we have $i(X)\omega = 0$, and therefore finally the formula

(2.18) $$\nabla^*_X \omega = i(X)d\omega$$

for the Bott connection on Q^*.

2.19 FOLIATED BUNDLES VERSUS NORMAL BUNDLES OF FOLIATIONS.

It is worth repeating that the normal bundle of a foliation is canonically equipped with the foliated structure described by the Bott connection. In contrast the foliation of a bundle over a foliated base space in general is a piece of geometric data given in addition. In this sense foliated bundles play with respect to the normal bundles of foliated manifolds the same role as that played by arbitrary bundles with respect to the tangent bundles of manifolds. It hardly needs to be elaborated that this point of view is appropriate as soon as functorial considerations play a role. This in particular is so in the discussion of characteristic classes, which is one of the reasons why we introduce the concept of foliated bundles. The other equally important reason is the wide range of geometric situations which can be successfully described with this concept. This is in particular so if one considers G-reductions of the normal bundle Q of a foliation.

2.20 CONSTRUCTION PRINCIPLES FOR FOLIATED BUNDLES. We discuss a few general ways how to get new foliated bundles out of old ones.

2.21 EXAMPLE. Let $\overline{P}$ be a foliated $\overline{G}$-bundle and $G \subset \overline{G}$ a closed subgroup. Then the G-bundle $\overline{P} \longrightarrow P = \overline{P}/G$ is canonically foliated. This generalizes the example 1.31 where it was explained how a connection in $\overline{P} \longrightarrow \overline{P}/\overline{G}$ induces a partial connection in $\overline{P} \longrightarrow \overline{P}/G$.

Let now $\overline{\omega}$ be a connection in the $\overline{G}$-bundle $\overline{P}$ adapted to a partial connection. Let θ be a G-equivariant splitting of the exact sequence

$$(2.22) \qquad 0 \longrightarrow \underline{\underline{g}} \overset{\theta}{\leftrightarrows} \overline{\underline{\underline{g}}} \longrightarrow \overline{\underline{\underline{g}}}/\underline{\underline{g}} \longrightarrow 0.$$

Then the $\underline{\underline{g}}$-valued 1-form $\omega = \theta \circ \overline{\omega}$ on $\overline{P}$ defines a connection on the G-bundle $\overline{P} \longrightarrow \overline{P}/G$ which is adapted to the canonically induced partial connection. The horizontal space of $\theta \circ \omega$ contains the vertical vectors on $\overline{G} \longrightarrow \overline{P} \longrightarrow P/\overline{G}$ which are tangent to the subspace of the fiber corresponding to $\ker \theta \cong \overline{\underline{\underline{g}}}/\underline{\underline{g}}$.

2.23 EXAMPLE. Let $f: M \longrightarrow M'$ be a submersion and $P' \longrightarrow M'$ any G-bundle. The pull-back $f^*P' = P$ is canonically foliated with respect to the foliation $T(f)$ on M given by the fibers of f. If $\overline{f}: P \longrightarrow P'$ denotes the canonical G-bundle map over $f: M \to M'$ then the subbundle $T(\overline{f}) \subset T_P$ is the involutive subbundle defining the partial flat connection.

Let $P' \longrightarrow M'$ be a foliated bundle and $f: M \longrightarrow M'$ a smooth map. The pull-back f^*L' of the involutive subbundle $L' \subset T_{M'}$ is an involutive subbundle $f^*L' \subset f^*T_{M'}$. For f^*L' to define a (non-singular) foliation of M we have to require that the map f is transversal to the foliation of M', i.e. the composition

$$T_x(M) \xrightarrow{df_x} T_{f(x)}M' \longrightarrow Q'_{f(x)}$$

is surjective for all $x \in M$, where $Q' = T_{M'}/L'$. In that case the pull-back $f^*P' \longrightarrow M$ is canonically foliated by the pull-back of the flat partial connection on $P' \longrightarrow M'$.

2.24 BASIC CONNECTIONS. The following special type of adapted connections is often useful.

DEFINITION. An adapted connection ω in a foliated principal bundle $P \longrightarrow M$ is basic, if

$$\theta(X)\omega = 0 \tag{2.25}$$

for the Lie derivative $\theta(X)$ of every partially horizontal vectorfield X on P.

The geometric significance of (2.25) is that the flows generated by partially horizontal vectorfields leave the connection form ω invariant.

The existence of a basic adapted connection is a special property of a foliated bundle, which may or may not hold. This condition is automatically satisfied for an ordinary bundle (with the foliation by points of the base space), since the zero vectorfield is the only partially horizontal vectorfield.

From the identity $\theta(X) = i(X)d + di(X)$ and the property $i(X)\omega = 0$ for a partially horizontal vectorfield X on P it follows [KT 7, (7.9)]

$$\theta(X)\omega = 0 \iff i(X)d\omega = 0 \iff i(X)\Omega = 0 \tag{2.26}$$

The last equivalence follows from the structure equation $\Omega = d\omega + \frac{1}{2}[\omega,\omega]$, since $i(X)\omega = 0$.

It follows in particular that a flat connection is always basic.

The following characterization will be useful.

2.27 PROPOSITION [KT 7]. Let $P \longrightarrow M$ be a foliated G-bundle, and ω an adapted connection. Then ω is basic if and only if for every $\alpha \in \underline{\underline{g}}^*$ we have $\alpha\Omega \in \Gamma(P,\pi^*\wedge^2 \underline{Q}^*.\Omega_P^0)$.

Proof. $\underline{Q}^*$ denotes as before the sheaf of local 1-forms on M annihilating the vectorfields belonging to the foliation $L \subset T_M$. We use the notations in the proof of proposition 1.43. Formula (1.45) implies that $\alpha\Omega$ is locally representable in the form

$$\alpha\Omega = \sum_{i<j} \alpha\Omega(\tilde{Y}_i,\tilde{Y}_j)\beta_i \wedge \beta_j + \sum_{i,j} \alpha\Omega(\tilde{X}_i,\tilde{Y}_j)\alpha_i \wedge \beta_j .$$

But (2.26) shows that ω is basic if and only if the second sum vanishes, i.e. if and only if

(2.28) $$\alpha\Omega = \sum_{i<j} \alpha\Omega(\tilde{Y}_i,\tilde{Y}_j)\beta_i \wedge \beta_j .$$

Since $\beta_1,\ldots,\beta_q$ form locally a frame of π^*Q^*, this establishes the desired result. □

Let $E \longrightarrow M$ be a vectorbundle associated to $P \longrightarrow M$ by a representation $G \longrightarrow GL(V)$, and ∇ the covariant derivative with curvature R defined by a basic connection ω.

2.29 LEMMA. For every vectorfield X on M belonging to the foliation $L \subset T_M$ we have

(2.30) $$i(X)R = 0.$$

Proof. Let X,Y be any vectorfields on M and $\tilde{X},\tilde{Y}$ their horizontal lifts with respect to ω. For a section s of E we consider the corresponding equivariant map $f : P \longrightarrow V$ and the relation (2.8) between the covariant derivative of s and the Lie derivative of f. From the formula for R in terms of ∇ it follows then immediately that

$$(2.31) \qquad (R(X,Y)s)(x) = u(([\tilde{X},\tilde{Y}] - \widetilde{[X,Y]})_u f)$$

Since $\tilde{X},\tilde{Y}$ are projectable vectorfields under $\pi : P \longrightarrow M$, we have $\pi[\tilde{X},\tilde{Y}] = [X,Y]$. It follows that $\widetilde{[X,Y]}_u$ is the horizontal component of $[\tilde{X},\tilde{Y}]_u$ and hence $([\tilde{X},\tilde{Y}] - \widetilde{[X,Y]})_u$ the vertical component of $[\tilde{X},\tilde{Y}]_u$. But the identity

$$\Omega(\tilde{X},\tilde{Y}) = d\omega(\tilde{X},\tilde{Y}) = -\omega[\tilde{X},\tilde{Y}]$$

shows that this component at $u \in P$ is also given by the value at $u \in P$ of the fundamental vectorfield defined by $-\Omega_u(\tilde{X},\tilde{Y}) \in \underline{\underline{g}}$. Therefore (2.31) implies the formula

$$(2.32) \qquad (R(X,Y)s)(u) = u((-\Omega_u(\tilde{X},\tilde{Y})^*f)(u))$$

The desired result follows now trivally from this formula and (2.26)□

We observe in passing that in view of (2.9) the formula (2.32) can also be written in the form

$$(R(X,Y)s)(u) = u(d\rho(\Omega_u(\tilde{X},\tilde{Y})).f(u))$$

where $d\rho(\Omega_u(\tilde{X},\tilde{Y})) \in \underline{\underline{gl}}(V)$ acts on $f(u) \in V$.

Lemma 2.29 justifies the following terminology.

2.33 DEFINITION. <u>An adapted connection</u> ∇ <u>in a foliated vector-bundle</u> $E \longrightarrow M$ <u>is basic, if</u>

(2.34) $$i(X)R = 0$$

<u>for every vectorfield</u> X <u>on</u> M <u>belonging to the foliation</u> $L \subset T_M$.

Note that $i(X)R$ is a 1-form on M with values in the bundle $\mathrm{End}(E)$. For a partial connection in E (along $L \subset T_M$) it is only defined on sections of L. To be defined on all vectorfields of M, we need an extension to an adapted connection in E.

Remark. The terms adapted and basic connection are not always used with these meanings by other authors. We have used these terms in all our papers on the subject in the sense explained above. A projectable connection in the sense of Molino [MO 1] is a basic connection in the sense used here.

2.35 EXAMPLES OF BASIC CONNECTIONS. The reason for the term basic is seen in the following example. Let $f : M \longrightarrow M'$ be a submersion, $P' \longrightarrow M'$ a principal G-bundle and $P = f^*P'$. By example 2.21 the bundle $P \longrightarrow M$ is canonically foliated with respect to the foliation of M by the fibers of f with tangent bundle $T(f)$. We claim that for any connection ω' in P' the pullback connection $\omega = \overline{f}^*\omega'$ ($\overline{f} : P \longrightarrow P'$ the induced bundle map) is an adapted connection which is basic. Since $T(\overline{f}) \subset T_P$ is the canonical foliation in P, it is clear that ω is adapted to this partial connection of P. A partially horizontal vectorfield X on P is a vectorfield belonging to $T(\overline{f})$. It is clear that the local flow generated by such a vectorfield leaves ω invariant, since ω is simply the pullback $f^*\omega'$. This translates infinitesimally into $\theta(X)\omega = 0$. Any foliated bundle is locally of the form of this example (using the local submersions defining the foliation of the base space). Therefore a connection in a foliated bundle is basic precisely if it is locally a pullback as described above.

From these remarks it is obvious that a basic connection exists in the normal bundle Q of a foliation, provided the foliation is given by local submersions onto a manifold, such that these submersions are compatible by local diffeomorphisms preserving a connection in the target manifold. Such a connection pulls back to a basic connection in Q. An example is a Riemannian foliation (see Pasternack [P 2]), which is a foliation defined by local submersions onto a Riemannian manifold, which are compatible by local isometries of the target Riemannian manifold. The Riemannian connection of the target manifold pulls back to a basic connection in the normal bundle Q of the Riemannian foliation.

2.36 EXAMPLE. Another situation where basic connections are certain to exist is the following. Let $K \times M \longrightarrow M$ be an almost free compact group action lifting to an action $\tau : K \longrightarrow \mathrm{Aut}(P)$ by bundle automorphisms of the G-bundle $P \longrightarrow M$ as in example 2.4. Then any adapted connection ω on P can be integrated over K with respect to the unit measure to give a basic connection form

$$\hat{\omega} = \int_K \tau^*(k)\omega \quad \text{on} \quad P. \tag{2.37}$$

2.38 More generally let $P \longrightarrow M$ be foliated, X a vectorfield belonging to $L \subset T_M$ and $\tilde{X}$ its partially horizontal lift. The flow $\tilde{\varphi}_t$ of $\tilde{X}$ acts by (local) bundle automorphisms on P which preserve the foliation. An adapted connection ω clearly is basic if and only if

$$\tilde{\varphi}_t^* \omega = \omega \tag{2.39}$$

for all such flows. The integration procedure of the preceding example can be applied as soon as the group of bundle automorphisms preserving the foliation is compact. In such a case the existence of a basic connection follows.

2.40 EXAMPLE. Consider as in example 2.21 a foliated $\overline{G}$-bundle $\overline{P}$, a closed subgroup $G \subset \overline{G}$ and the foliated G-bundle $\overline{P} \longrightarrow \overline{P}/G$. Let $\overline{\omega}$ be a basic connection in the foliated $\overline{G}$-bundle $\overline{P}$, $\theta : \overline{\mathfrak{g}} \longrightarrow \mathfrak{g}$ a G-equivariant splitting of $0 \to \mathfrak{g} \to \overline{\mathfrak{g}} \to \overline{\mathfrak{g}}/\mathfrak{g} \to 0$ and $\omega = \theta \circ \overline{\omega}$ the corresponding adapted connection in the G-bundle $\overline{P}$ (see 2.31).

2.41 LEMMA. <u>For every basic connection in the foliated</u> $\overline{G}$-<u>bundle</u> $\overline{P} \longrightarrow \overline{P}/\overline{G}$ <u>the connection</u> $\omega = \theta \circ \overline{\omega}$ <u>in the foliated</u> G-<u>bundle</u> $\overline{P} \longrightarrow \overline{P}/G$ <u>is basic</u>.

Proof. Let $\overline{\Omega}$ be the curvature of $\overline{\omega}$ on the $\overline{G}$-bundle $\overline{P} \longrightarrow \overline{P}/\overline{G}$ and Ω the curvature of ω on the G-bundle $\overline{P} \longrightarrow \overline{P}/G$. Then

$$\Omega = d\omega + \frac{1}{2}[\omega,\omega] = d(\theta \circ \overline{\omega}) + \frac{1}{2}[\theta \circ \overline{\omega}, \theta \circ \overline{\omega}]$$

$$= \theta \circ \overline{\Omega} + \frac{1}{2}([\theta \circ \overline{\omega}, \theta \circ \overline{\omega}] - \theta[\overline{\omega},\overline{\omega}])$$

To show that $i(X)\Omega = 0$ for partially horizontal vectorfields on the G-bundle P, it suffices to consider the following two cases: (i) X is also partially horizontal with respect to $\overline{\omega}$ on the $\overline{G}$-bundle $\overline{P}$, and (ii) X is vertical on the $\overline{G}$-bundle $\overline{P}$. But in both cases $i(X)$ annihilates the right hand expression. Case (i) is obvious and case (ii) follows from $i(X)\overline{\Omega} = 0$ (since X is vertical) and $\theta \circ \overline{\omega}(X) = 0$ (since $\ker \theta \cong \overline{\mathfrak{g}}/\mathfrak{g}$). Therefore $i(X)\Omega = 0$ in both cases and ω is indeed basic. □

2.42 COROLLARY. <u>Let</u> $\overline{P}$ <u>be a</u> $\overline{G}$-<u>bundle with flat connection</u> $\overline{\omega}$, $G \subset \overline{G}$ <u>a closed subgroup and</u> $\theta : \overline{\mathfrak{g}} \longrightarrow \mathfrak{g}$ <u>a</u> G-<u>equivariant splitting</u>

of $0 \to \underline{\underline{g}} \to \overline{\underline{\underline{g}}} \to \overline{\underline{\underline{g}}}/\underline{\underline{g}} \to 0$. *Then* $\omega = \theta \circ \overline{\omega}$ *is a basic connection in the foliated* G-*bundle* $\overline{P} \to \overline{P}/G$.

2.43 OBSTRUCTION FOR BASIC CONNECTIONS. We explain in this section how the existence of a basic connection is characterized by the vanishing of a certain 1-dimensional cohomology class ([KT 2,3] and [MO 1]). For this it is useful first to interprete a connection in the bundle $\pi : P \to M$ as a splitting ω of the exact sequence of $\underline{O}_M$-modules

$$\underline{A}(P) : 0 \to \Omega^1_M \to \pi^G_* \Omega^1_P \to \underline{P}(\underline{\underline{g}}^*) \to 0 \quad (\omega : \underline{P}(\underline{\underline{g}}^*) \to \pi^G_* \Omega^1_P)$$

as first explained by Atiyah [AT]. Here $\underline{P}(\underline{\underline{g}}^*)$ denotes the sheaf of sections of the vectorbundle $P \times_G \underline{\underline{g}}^*$ on M associated to P by the adjoint representation. π_* denotes the direct image functor and $\pi^G_* \Omega^1_P$ the subsheaf of G-invariant 1-forms in $\pi_* \Omega^1_P$. The exact sequence $\underline{A}(P)$ can be thought of as an element $\zeta(\underline{A}(P))$ of a sheaf cohomology group $H^1(M, \underline{Hom}_O(\underline{P}(\underline{\underline{g}}^*), \Omega^1_M))$ in the fashion standard in homological algebra (see [KT 3]). There is no point to this in the smooth case, since this cohomology group is trivial and the corresponding extensions are trivial.

Now if $\pi : P \to M$ is equipped with a foliation on M characterized by an $\underline{O}_M$-submodule $\underline{Q}^* \subset \Omega^1_M$, a foliation of P is a splitting ω_0 of the exact sequence of $\underline{O}_M$-modules

$$\lambda_* \underline{A}(P) : 0 \to \Omega^1_M/\underline{Q}^* \to \pi^G_* \Omega^1_P/\underline{Q}^* \to \underline{P}(\underline{\underline{g}}^*) \to 0 \quad (\omega_0 : \underline{P}(\underline{\underline{g}}^*) \to \pi^G_* \Omega^1_P/\underline{Q}^*)$$

where $\lambda : \Omega^1_M \to \Omega^1_M/\underline{Q}^*$ is the canonical map (see [KT 7], (1.4) (1.5)).

A connection ω adapted to the foliation defined by ω_0 is then a splitting of the exact sequence of $\underline{O}_M$-modules (see [KT 7], p. 27/28)

$$\underline{A}(P, \omega_0) : 0 \to \underline{Q}^* \to \pi^G_* \tilde{\Omega} \to \underline{P}(\underline{\underline{g}}^*) \to 0$$

where $\tilde{\Omega} \subset \Omega^1_P$ is a submodule defined by ω_0. Again this sequence can be thought of as an element $\zeta(\underline{A}(P,\omega_0))$ in the sheaf cohomology group $H^1(M,\underline{Hom}_O(\underline{P}(\underline{g}^*),\underline{Q}^*))$. Whereas this group is zero in the case of a smooth non-singular foliation, it can be non-trivial for a singular foliation even in the smooth case, which is one reason why this approach is of interest.

To return now to the problem of the existence of a basic connection, we first observe that all terms of the sequence $\underline{A}(P,\omega_0)$ carry a canonical action by the involutive sheaf $\underline{L} = \underline{Ann}(\underline{Q}^*)$. An adapted connection is basic if and only if it is a splitting of $\underline{A}(P,\omega_0)$ compatible with these $\underline{L}$-actions. In the parlance of [KT 3] the sequence $\underline{A}(P,\omega_0)$ is a sequence of $\underline{U}(\underline{L},\underline{O})$-modules and a basic connection in P a split of $\underline{A}(P,\omega_0)$ as a sequence of $\underline{U}(\underline{L},\underline{O})$-modules. The sequence $\underline{A}(P,\omega_0)$ can again in standard fashion be interpreted as an element $\zeta(\underline{A}(P,\omega_0))$ in a sheaf cohomology group $H^1(M,\underline{L};\underline{Hom}_O(\underline{P}(\underline{g}^*),\underline{Q}^*))$. This group need not be trivial even in the smooth case and $\zeta(\underline{A}(P,\omega_0))$ is the obstruction to the existence of a basic connection in P. It has been shown in [KT 6,7] how this cohomology class is related to the generalized characteristic classes of the foliated bundle P defined in chapter 4.

2.44 HOMOGENEOUS FOLIATED BUNDLES [KT 9,10]. Let $H \subset G \subset \overline{G}$ be subgroups with H closed in $\overline{G}$. Consider the G-bundle

$$P = \overline{G} \times_H G \longrightarrow \overline{G}/H \tag{2.45}$$

It carries a canonical foliation given as follows. The formula

$$(\overline{g},g)\cdot g' = (\overline{g}g', g'^{-1}g) \tag{2.46}$$

for $(\overline{g},g) \in \overline{G} \times G$ and $g' \in G$ defines the diagonal right action of G on $\overline{G} \times G$. The G-orbits define a foliation on $\overline{G} \times G$, which under the projection $\overline{G} \times G \longrightarrow \overline{G}$ maps onto the coset foliation of $\overline{G}$ given by the right action of G on $\overline{G}$. Since the projection is G-equivariant, the bundle P in (2.45) inherits a foliated bundle structure. We call such a foliated bundle homogeneous.

The interest of the homogeneous foliated bundle P is that the normal bundle Q_G of the homogeneous foliation on $\bar{G}/H$ given by G (see 1.28) is associated to P, and the canonical foliation on Q_G is inherited from the canonical foliation of P. To see this, consider the exact sequence of H-modules

(2.47) $$0 \longrightarrow \underline{\underline{g}}/\underline{\underline{h}} \longrightarrow \bar{\underline{\underline{g}}}/\underline{\underline{h}} \longrightarrow \bar{\underline{\underline{g}}}/\underline{\underline{g}} \longrightarrow 0$$

and the corresponding exact sequence of vectorbundles on $\bar{G}/H$ associated to the principal bundle $H \longrightarrow \bar{G} \longrightarrow \bar{G}/H$

(2.48) $$0 \longrightarrow \bar{G} \times_H \underline{\underline{g}}/\underline{\underline{h}} \longrightarrow \bar{G} \times_H \bar{\underline{\underline{g}}}/\underline{\underline{h}} \longrightarrow \bar{G} \times_H \bar{\underline{\underline{g}}}/\underline{\underline{g}} \longrightarrow 0$$

The middle term is $T_{\bar{G}/H}$, and the first term precisely the tangent bundle L_G of the G-foliation on $\bar{G}/H$, so that the third term equals $Q_G = T_{\bar{G}/H}/L_G$. But since

$$\bar{G} \times_H \bar{\underline{\underline{g}}}/\underline{\underline{g}} \equiv (\bar{G} \times_H G) \times_G \bar{\underline{\underline{g}}}/\underline{\underline{g}}$$

it follows that

(2.49) $$Q_G \cong P \times_G \bar{\underline{\underline{g}}}/\underline{\underline{g}}.$$

It is now easy to verify that the foliation on Q_G induced from P coincides with the Bott connection (2.15) in Q_G.

2.50 LOCALLY HOMOGENEOUS FOLIATED BUNDLES [KT 9,10]. Let $\Gamma \subset \bar{G}$ be in addition a discrete subgroup operating properly discontinuously and without fixed points on $\bar{G}/H$, so that the double coset space $\Gamma\backslash\bar{G}/H$ is a manifold. Then the previous discussion generalizes to the G-bundle

$$(2.51) \qquad P = (\Gamma\backslash\overline{G}) \times_H G \longrightarrow \Gamma\backslash\overline{G}/H$$

which inherits a canonical foliated structure, since the foliated structure on (2.45) is obviously invariant under the left action of Γ. The normal bundle of the locally homogeneous foliation by G on $\Gamma\backslash\overline{G}/H$ (see 1.29) is then again associated to the locally homogeneous foliated bundle.

3. CHARACTERISTIC CLASSES OF FLAT BUNDLES.

Let $P \longrightarrow M$ be a principal bundle with structure group G. The simplest non-trivial case of a foliation of P is a flat connection ω on P. In this chapter we discuss the characteristic classes arising from the existence of ω [KT 1, 5 to 10]. This construction is itself of geometric interest, as the examples in chapter 6 will show. It further serves as a motivation for the general construction of characteristic classes for foliated bundles in chapter 4. Some facts are stated in this chapter without proof, since they will be established later in a more general context.

For every $\alpha \in \underline{\underline{g}}^*$ we have a 1-form $\alpha\omega \in \Omega^1(P)$. This defines a linear map $\wedge^1\underline{\underline{g}}^* \cong \underline{\underline{g}}^* \longrightarrow \Omega^1(P)$ and by multiplicative extension an algebra homomorphism

$$\wedge^{\cdot}\underline{\underline{g}}^* \xrightarrow{\omega} \Omega^{\cdot}(P) \tag{3.1}$$

The exterior algebra $\wedge^{\cdot}\underline{\underline{g}}^*$ is equipped with the Chevalley-Eilenberg differential $d = d_\wedge$

$$d\varphi(x_1,\ldots,x_{q+1}) = \sum_{i<j} (-1)^{i+j}\varphi([x_i,x_j],x_1,\ldots,\hat{x}_i,\ldots,\hat{x}_j,\ldots,x_{q+1}) \tag{3.2}$$

for $\varphi \in \wedge^q\underline{\underline{g}}^*$ and $x_1,\ldots,x_{q+1} \in \underline{\underline{g}}$. The algebra $\Omega^{\cdot}(P)$ of differential forms on P is equipped with the exterior differential $d = d_P$.

3.3 PROPOSITION. *The homomorphism (3.1) is a homomorphism of differential algebras, i.e. commutes with the differentials on* $\wedge^{\cdot}\underline{\underline{g}}^*$ *and* $\Omega^{\cdot}(P)$.

More generally one can consider for any connection ω in P the expression

(3.4) $$d_P\omega - \omega d_\wedge : \underline{\underline{g}}^* \longrightarrow \Omega^2(P)$$

and prove the following result, implying (3.3).

3.5 PROPOSITION. *Let* Ω *be the curvature of a connection* ω *in* P, *identified with the map* $\underline{\underline{g}}^* \longrightarrow \Omega^2(P)$ *defined by* $\alpha \to \Omega(\alpha) = \alpha\Omega$. *Then*

$$\Omega(\alpha) = (d_P\omega - \omega d_\wedge)(\alpha) \quad \textit{for any} \quad \alpha \in \underline{\underline{g}}^*.$$

Proof. The structure equation implies for each $\alpha \in g^*$

$$\Omega(\alpha) \equiv \alpha\Omega = \alpha d\omega + \frac{1}{2}\alpha[\omega,\omega]$$

Clearly with our notations

$$\alpha d\omega = d_P\alpha\omega = d_P\omega\alpha$$

We therefore have to establish the identity

(3.6) $$\frac{1}{2}\alpha[\omega,\omega] = -\omega d_\wedge \alpha \quad \text{for} \quad \alpha \in \underline{\underline{g}}^*.$$

To evaluate the left hand side

$$\frac{1}{2}\alpha[\omega,\omega](Y,Z) = \frac{1}{2}\alpha([\omega(Y),\omega(Z)] - [\omega(Z),\omega(Y)])$$

for vectorfields Y,Z on P, let $x_1,\dots,x_m$ be a basis of $\underline{\underline{g}}$ and $x_1^*,\dots,x_m^*$ a dual basis of $\underline{\underline{g}}^*$. Then in $\underline{\underline{g}}$ we have

$$\omega(Y) = \sum_j (x_j^*\,\omega(Y)).x_j$$

$$\omega(Z) = \sum_j (x_j^*\,\omega(Z)).x_j$$

Let further $\theta(x) : \underline{\underline{g}}^* \longrightarrow \underline{\underline{g}}^*$ be the contragredient reprepresentation of the adjoint representation, i.e.

(3.7) $$\theta(x)\alpha = -(\mathrm{ad}\ x)^*\alpha \quad \text{for} \quad \alpha \in \underline{\underline{g}}^*,\ x \in \underline{\underline{g}}$$

where $(\mathrm{ad}\ x)(y) = [x,y]$ for $y \in \underline{\underline{g}}$. In other words

(3.8) $$(\theta(x)\alpha)(y) = -\alpha[x,y]$$

Then e.g.

$$\alpha[\omega(Y),\omega(Z)] = \alpha[\sum_j (x_j^*\ \omega(Y)).x_j,\omega(Z)]$$

$$= -\sum_j (x_j^*\ \omega(Y)).(\theta(x_j)\alpha)(\omega(Z)) = -\sum_j (\omega\ x_j^*)(Y).\omega(\theta(x_j)\alpha)(Z)$$

and similarly for $\alpha[\omega(Z),\omega(Y)]$. It follows that

$$\tfrac{1}{2}\ \alpha[\omega,\omega](Y,Z)$$

$$= \tfrac{1}{2}\left\{-\sum_j (\omega\ x_j^*)(Y)\ .\ \omega(\theta(x_j)\alpha)(Z) + \sum_j (\omega\ x_j^*)(Z)\ .\ \omega(\theta(x_j)\alpha)(Y)\right\}$$

$$= -\tfrac{1}{2}\sum_j (\omega\ x_j^* \wedge \omega(\theta(x_j)\alpha))(Y,Z)$$

$$= -\tfrac{1}{2}\sum_j \omega(x_j^* \wedge \theta(x_j^*)\alpha)(Y,Z)$$

since ω is a multiplicative homomorphism. To establish (3.6) it suffices therefore to verify the identity

(3.9) $$d_\wedge\alpha = \tfrac{1}{2}\sum_j x_j^* \wedge \theta(x_j)\alpha$$

for the Chevalley-Eilenberg differential $d_\wedge$ applied to $\alpha \in \underline{\underline{g}}^*$. This is a classical formula of Koszul [K 1], which is verified as follows. Note first that by (3.2) (3.8) for $y,z \in \underline{\underline{g}}$

$$(d_\wedge \alpha)(y,z) = -\alpha[y,z] = (\theta(y)\alpha)(z)$$

It suffices therefore to check

$$(\theta(y)\alpha)(z) = \frac{1}{2} \sum_j (x_j^* \wedge \theta(x_j)\alpha)(y,z)$$

on basis elements $y = x_k$, $z = x_\ell$. For the right hand side we obtain then

$$\frac{1}{2} \sum_j \{x_j^*(x_k).(\theta(x_j)\alpha)(x_\ell) - x_j^*(x_\ell).(\theta(x_j)\alpha)(x_k)\}$$

$$= \frac{1}{2} ((\theta(x_k)\alpha)(x_\ell) - (\theta(x_\ell)\alpha)(x_k)) = -\alpha[x_k,x_\ell],$$

the same as for the left hand side

$$(\theta(x_k)\alpha)(x_\ell) = -\alpha[x_k,x_\ell].$$

This proves Koszul's formula (3.9) and finishes the proof of proposition 3.5. □

The Lie algebra cohomology is by definition $H^\cdot(\underline{\underline{g}}) \equiv H(\wedge^\cdot \underline{\underline{g}}^*, d_\wedge)$. The DeRham cohomology $H^\cdot_{DR}(P)$ is the cohomology of the complex $(\Omega^\cdot(P), d_P)$. From (3.3) we have the following consequence.

3.10 COROLLARY. <u>Let</u> ω <u>be a flat connection in</u> P. <u>Then the canonical</u> DG-<u>homomorphism</u> $\omega : \wedge^\cdot \underline{\underline{g}}^* \longrightarrow \Omega^\cdot(P)$ <u>induces a cohomology map</u>

$$k_* : H^\cdot(\underline{\underline{g}}) \longrightarrow H^\cdot_{DR}(P).$$

The simplest example of Corollary 3.10 is the group G itself viewed as principal fibration over a point. The canonical connection form ω is then the $\underline{\underline{g}}$-valued Maurer-Cartan form of G. The corresponding DG-homomorphism $\wedge^\cdot \underline{\underline{g}}^* \longrightarrow \Omega^\cdot(G)$ is the canonical inclusion of the left-invariant forms into $\Omega^{\cdot}(G)$, which induces a canonical homomorphism $H^{\cdot}(\underline{\underline{g}}) \longrightarrow H^{\cdot}_{DR}(G)$. For compact connected G this is an isomorphism, since every DeRham cohomology class can be represented by an invariant form.

To obtain cohomology classes in the base space M of a flat bundle $P \longrightarrow M$ one can use several devices. A lengthy formal digression is desirable before we explain such a procedure. An impatient reader may wish to first see this procedure in Theorem 3.30 before digesting the algebraic jargon developed in 3.11 and the paragraphs following it.

But before embarking on this digression, we wish to discuss the cohomology classes on P defined by k_* (in 3.10) on the cochain level. Let $\underline{\Psi} \in \Lambda^q \underline{\underline{g}}^*$ be a cycle and ω the $\underline{\underline{g}}$-valued connection 1-form on P. Then $k_*[\Psi]$ is represented by the q-form on P

$$\Psi(\underbrace{\omega \wedge \ldots \wedge \omega}_{q \text{ factors}})$$

where $\omega \wedge \ldots \wedge \omega$ is the $\Lambda^q \underline{\underline{g}}$ - valued q-th exterior product of ω with itself with respect to the multiplication in $\Lambda^{\cdot}\underline{\underline{g}}$.

At this point it will be useful to recall that for forms with values in a graded (associative) algebra $E^{\cdot}$ the exterior product of an E^r-valued p-form φ and an E^s-valued q form Ψ is the E^{r+s}-valued (p+q)-form

$$(\varphi \wedge \Psi)(X_1,\ldots,X_{p+q}) = \sum_{\sigma} \varepsilon_\sigma \, \varphi(X_{\sigma(1)},\ldots,X_{\sigma(p)}) \cdot \Psi(X_{\sigma(p+1)},\ldots,X_{\sigma(p+q)}).$$

The summation extends over all permutations σ of $(1,2,\ldots,p+q)$ such that $\sigma(1) < \ldots < \sigma(p)$ and $\sigma(p+1) < \ldots < \sigma(p+q)$, and ε_σ

is the signature of σ. If $E^\cdot$ is graded commutative, i.e.

$$a.a' = (-1)^{\deg a.\deg a'} a'.a$$

then we have the commutation rule

$$\varphi \wedge \Psi = (-1)^{pq+rs} \Psi \wedge \varphi.$$

Thus e.g. for two $\underline{\underline{g}}$-valued 1-forms φ, Ψ their exterior product with respect to the multiplication in $\wedge^\cdot \underline{\underline{g}}$ satisfies the commutation rule

$$\varphi \wedge \Psi = \Psi \wedge \varphi.$$

For clarity's sake we recall that for forms with values in a graded Lie algebra $E^\cdot$ the exterior product of an E^r-valued p-form and an E^s-valued q-form is denoted $[\varphi,\Psi]$ (or sometimes $[\varphi \wedge \Psi]$). The commutation rule

$$[a,a'] = (-1)^{\deg a.\deg a'+1}[a',a]$$

in $E^\cdot$ implies the commutation rule

$$[\varphi,\Psi] = (-1)^{pq+rs+1}[\Psi,\varphi]$$

for $E^\cdot$-valued forms. For 1-forms φ, Ψ with values in an ungraded Lie algebra $\underline{\underline{g}}$ e.g.

$$[\varphi,\Psi](X,Y) = [\varphi(X),\Psi(Y)] - [\varphi(Y),\Psi(X)]$$

with the commutation rule

$$[\varphi,\Psi] = [\Psi,\varphi].$$

3.11 G-DG-ALGEBRAS AND $\underline{\underline{g}}$-DG-ALGEBRAS. For several purposes useful in this work we describe algebraic concepts introduced by Cartan, Koszul and Weil in [CA] (see also the exposition in [GHV, Vol. III]).

These concepts have been used extensively by the authors in their work [KT, 5 to 12].

Recall first that a DG-algebra $E^{\cdot}$ is a differential (positively) graded algebra (over the reals), i.e. a graded algebra equipped with a differential $d = d_E$ which is a derivation of degree 1. A derivation $D : E^{\cdot} \longrightarrow E^{\cdot+r}$ of degree r is an $\mathbb{R}$-linear homomorphism of degree r such that

$$D(a.a') = Da.a' + (-1)^{\deg a.r} a.Da'$$

The commutator

$$[D,D'] = DD' + (-1)^{rr'+1} D'D$$

of derivations D and D' of degree r and r' is a derivation of degree $r+r'$.

3.12 DEFINITION. Let E be a DG-algebra and G a Lie group with Lie algebra $\underline{\underline{g}}$. Then E is a G-DG-algebra, if it is equipped with derivations $i(x) : E^{\cdot} \longrightarrow E^{\cdot-1}$ of degree -1 for every $x \in \underline{\underline{g}}$, and DG-algebra automorphisms $\rho(g) : E \longrightarrow E$ for every $g \in G$ such that the following condition hold:

(i) $i(x)^2 = 0$ for $x \in \underline{\underline{g}}$;

(ii) $\rho(gg') = \rho(g)\rho(g')$ for $g,g' \in G$;

(iii) $\rho(g)i(y)\rho(g^{-1}) = i(\mathrm{Ad}(g)y)$ for $y \in \underline{\underline{g}}$, $g \in G$;

(iv) $\theta(x) = i(x)d + d\,i(x)$ for $x \in \underline{\underline{g}}$ and the differential $\theta = d\rho$ of ρ.

The differential $d\rho = \theta$ of the group homomorphism ρ is defined by

$$\theta(x) = \frac{d}{dt}\Big|_{t=0} \rho(\mathrm{expt}\ x) \quad \text{on} \quad x \in \underline{\underline{g}}.$$

It is a Lie algebra homomorphism

$$\theta : \underline{\underline{g}} \longrightarrow \text{Der } E$$

into the derivations of degree 0 of E equipped with the commutator bracket, i.e.

$$\theta[x,y] = [\theta(x),\theta(y)] \quad \text{for} \quad x,y \in \underline{\underline{g}}.$$

Differentiating (iii) for $g_t = \exp t\, x$ we obtain for $t = 0$

$$[\theta(x),\, i(y)] = i[x,y] \quad \text{for} \quad x,y \in \underline{\underline{g}}.$$

Thus E is a $\underline{\underline{g}}$-DG-algebra in the sense of the following definition.

3.13 DEFINITION. Let $E^{\cdot}$ be a DG-algebra and $\underline{\underline{g}}$ a Lie algebra. Then $E^{\cdot}$ is a $\underline{\underline{g}}$-DG-algebra, if it is equipped with derivations $i(x) : E^{\cdot} \longrightarrow E^{\cdot -1}$ of degree -1 and derivations $\theta(x) : E \longrightarrow E$ of degree 0 for every $x \in \underline{\underline{g}}$, such that the following conditions hold:

(i) $i(x)^2 = 0$ for all $x \in \underline{\underline{g}}$;

(ii) $\theta[x,y] = [\theta(x),\theta(y)]$ for all $x,y \in \underline{\underline{g}}$;

(iii) $[\theta(x),i(y)] = i[x,y]$ for all $x,y \in \underline{\underline{g}}$;

(iv) $\theta(x) = i(x)d + di(x)$ for $x \in \underline{\underline{g}}$.

Note that (iv) implies the formula

$$\theta(x)d = d\, \theta(x).$$

In fact (ii) is already a consequence of (i) (iii) (iv), but such remarks on minimal axioms are of no significance for our purpose. What we wish to have is a concise language, and useful properties, which are easy to verify in examples.

A G-DG-algebra is clearly a $\underline{\underline{g}}$-DG-algebra. For a $\underline{\underline{g}}$-DG-algebra E the representation $\theta : \underline{\underline{g}} \longrightarrow \text{Der } E$ may not necessarily be integrable to a group representation $\rho : G \longrightarrow \mathrm{Aut}\, E$. This is certainly so if G is simply connected, otherwise the usual difficulties occur, even if G is connected.

The geometric model for the concept of a G-DG-algebra is the De Rham complex $\Omega^{\cdot}(P)$ of differential forms on a principal G-bundle $P \longrightarrow M$. The fundamental vectorfield X^* on P associated to $x \in \underline{\underline{g}}$ is given by

$$X^*_u = \frac{d}{dt}\Big|_{t=0} (u \,.\, \mathrm{expt}\; x) \quad \text{at} \quad u \in P.$$

The definitions

$$(3.14) \qquad \begin{cases} i(x)\varphi = i(X^*)\varphi \\ \rho(g)\varphi = R^*_g\,\varphi \end{cases}$$

for $\varphi \in \Omega^q(P)$ turn $\Omega^{\cdot}(P)$ into a G-DG-algebra. The differential $\theta = d\rho$ is given by

$$(3.15) \qquad \theta(x)\varphi = \frac{d}{dt}\Big|_{t=0} R^*_{\mathrm{expt}x}\,\varphi = \theta(X^*)\varphi\;.$$

For the special case of the trivial fibration of G over a point, the fundamental vectorfield on G associated to $x \in \underline{\underline{g}}$ under the right action of G is the left invariant vectorfield corresponding to $x \in \underline{\underline{g}}$ and thus can be identified with x. Formulas (3.14) (3.15) read then for 1-forms $\alpha \in \Omega^1(G)$ and $x \in \underline{\underline{g}}$, $g \in G$ as follows:

$$(3.16) \qquad \begin{cases} i(x)\alpha = \alpha(x) \\ \rho(g)\alpha = \mathrm{Ad}(g^{-1})^*\alpha \end{cases}$$

and

$$(3.17) \qquad \theta(x)\alpha = -(\mathrm{ad}\; x)^*\alpha.$$

The complex of left invariant forms on G can be identified with $\wedge^{\cdot}\underline{\underline{g}}^*$. The exterior differential on $\Omega^{\cdot}(G)$ restricts on $\wedge^{\cdot}\underline{\underline{g}}^*$ to the Chevalley-Eilenberg differential (3.2). There exist unique derivations $i(x)$ of degree -1 and $\theta(x)$ of degree 0 characterized by the formulas

$$i(x)\alpha = \alpha(x)$$

(3.18)

$$\theta(x)\alpha = -(\mathrm{ad}\ x)^*\alpha$$

for $\alpha \in \wedge^1\underline{\underline{g}}^*$, $x \in \underline{\underline{g}}$ and $(\mathrm{ad}\ x)(y) = [x,y]$ for $y \in \underline{\underline{g}}$. Thus $\wedge^{\cdot}\underline{\underline{g}}^*$ is a $\underline{\underline{g}}$-DG-algebra, and in fact a G-DG-algebra.

Let $H \subset G$ be a Lie subgroup of G with Lie algebra $\underline{\underline{h}} \subset \underline{\underline{g}}$ and E a G-DG-algebra.

3.19 DEFINITION. The elements invariant under the representation ρ/H form the invariant subalgebra

$$E^H = \{a \in E \mid \rho(h)a = a \text{ for all } h \in H\}$$

The subalgebra of H-basic elements is given by

$$E_H = \{a \in E^H \mid i(x)a = 0 \text{ for all } x \in \underline{\underline{h}}\}.$$

Both are subcomplexes of E, as follows from the formulas in (3.12) and $E_H \subset E^H \subset E$.

For a Lie subalgebra $\underline{\underline{h}} \subset \underline{\underline{g}}$ and a $\underline{\underline{g}}$-DG-algebra E one defines similarly

$$E^{\underline{\underline{h}}} = \{a \in E \mid \theta(x)a = 0 \text{ for all } x \in \underline{\underline{h}}\}$$

(3.20)

$$E_{\underline{\underline{h}}} = \{a \in E^{\underline{\underline{h}}} \mid i(x)a = 0 \text{ for all } x \in \underline{\underline{h}}\}$$

To explain the terminology, consider again the G-DG-algebra $\Omega^{\cdot}(P)$ for a principal G-bundle $P \longrightarrow M$.

3.21 PROPOSITION. $\Omega^{\cdot}(P)_G \cong \Omega^{\cdot}(M)$.

Proof. Let $\varphi \in \Omega^q(P)$. Then $R_g^* \varphi = \varphi$ for all $g \in G$ means that φ is G-invariant and $i(x)\varphi = 0$ for all $x \in \underline{g}$ means that φ is a horizontal form. Thus φ is the lift of a form $\psi \in \Omega^q(M)$. Since $\pi : P \longrightarrow M$ induces an injective map $\pi^* : \Omega^{\cdot}(M) \longrightarrow \Omega^{\cdot}(P)$, the result follows. In fact we have

$$(3.22) \qquad \pi^* : \Omega^{\cdot}(M) \xrightarrow[\cong]{} \Omega^{\cdot}(P)_G \subset \Omega^{\cdot}(P) \qquad \square$$

More generally, for any closed subgroup $H \subset G$, the isomorphism (3.21) generalizes to the isomorphism

$$(3.23) \qquad \Omega^{\cdot}(P)_H \cong \Omega^{\cdot}(P/H)$$

where P/H denotes the orbitspace of P under the action of H. In fact the projection of the H-bundle $P \xrightarrow{\hat{\pi}} P/H$ induces an isomorphism

$$(3.24) \qquad \hat{\pi}^* : \Omega^{\cdot}(P/H) \xrightarrow{\cong} \Omega^{\cdot}(P)_H \subset \Omega^{\cdot}(P)$$

An example is the H-bundle $G \longrightarrow G/H$ itself, where therefore

$$(3.25) \qquad \Omega^{\cdot}(G/H) \cong \Omega^{\cdot}(G)_H \subset \Omega^{\cdot}(G)$$

Note that in this case G acts on the left on these complexes. For the invariant elements under the left G-action we have then

$$(3.26) \qquad {}^G\Omega^{\cdot}(G/H) \cong {}^G\Omega^{\cdot}(G)_H \subset {}^G\Omega^{\cdot}(G) \cong \Lambda \underline{g}^*$$

In this way the G-invariant forms on the homogeneous space G/H appear as a subcomplex of the Chevalley-Eilenberg complex of $\underline{g}$.

Note that in fact

$$(3.27)\qquad \begin{array}{ccc} {}^{G}\Omega^{\cdot}(G/H) & \cong & {}^{G}\Omega^{\cdot}(G)_{H} \subset \\ \uparrow\cong & & \uparrow\cong \quad\quad \Lambda\underline{\underline{g}}^{*} \\ \Lambda(\underline{\underline{g}}/\underline{\underline{h}})^{*H} & \cong & (\Lambda\underline{\underline{g}}^{*})_{H} \subset \end{array}$$

Before we end this digression, we recall the definition of the relative Lie algebra cohomology

$$(3.28)\qquad H^{\cdot}(\underline{\underline{g}},H) \equiv H(\Lambda^{\cdot}(\underline{\underline{g}}/\underline{\underline{h}})^{*H}) \cong H((\Lambda^{\cdot}\underline{\underline{g}}^{*})_{H})$$

We return now to a flat bundle $P \longrightarrow M$ and its canonical homomorphism $k_{*} : H^{\cdot}(\underline{\underline{g}}) \longrightarrow H^{\cdot}_{DR}(P)$. We describe a systematic procedure for producing invariants in the base space M.

Let $H \subset G$ be a closed subgroup and P' an H-reduction of P, i.e. there is an isomorphism $P' \times_{H} G \cong P$ of the extension of P' to a G-bundle. Such a reduction P' is given by a section $s : M \longrightarrow P/H$ of the map $P/H \longrightarrow M$ induced by $P \longrightarrow M$ in the form $P' = s^{*}P$ (as H-bundles), or in diagram form

$$(3.29)\qquad \begin{array}{ccc} H & = & H \\ \downarrow & & \downarrow \\ P' = s^{*}P & \longrightarrow & P \\ \downarrow & & \downarrow \\ M & \xrightarrow{s} & P/H \end{array}$$

3.30 THEOREM [KT 6,7]. <u>Let</u> $\pi : P \longrightarrow M$ <u>be a flat principal</u> G-<u>bundle</u>, $H \subset G$ <u>a closed subgroup and</u> P' <u>an</u> H-<u>reduction of</u> P <u>given by a section</u> $s : M \longrightarrow P/H$ <u>of the induced map</u> $\hat{\pi} : P/H \to M$. <u>Then there is a well-defined multiplicative homomorphism</u>

$$\Delta_{*} : H^{\cdot}(\underline{\underline{g}},H) \longrightarrow H^{\cdot}_{DR}(M)$$

Δ_{*} <u>is the generalized characteristic homomorphism of</u> P.

Proof. The construction of Δ_* is as follows. For a flat connection ω in P we have by (3.3) the DG-homomorphism

$$\omega : \wedge^{\cdot}\underline{\underline{g}}^* \longrightarrow \Omega^{\cdot}(P) \tag{3.31}$$

We verify that ω is in fact a homomorphism of G-DG-algebras. The formulas

$$i(x)\omega = x, \qquad x \in \underline{\underline{g}}$$

$$R_g^*\omega = \mathrm{Ad}(g^{-1})\omega, \quad g \in G$$

for the $\underline{\underline{g}}$-valued 1-form ω translate for ω viewed as a map $\omega : \wedge^1\underline{\underline{g}}^* \longrightarrow \Omega^1(P)$ into

$$i(x)\omega(\alpha) = i(x)\alpha, \quad x \in \underline{\underline{g}} \text{ and } \alpha \in \underline{\underline{g}}^* .$$

$$R_g^*\omega(\alpha) = \omega(\mathrm{Ad}(g^{-1})^*\alpha)$$

For the unique multiplicative extension (still denoted with the same symbol) $\omega : \wedge^{\cdot}\underline{\underline{g}}^* \longrightarrow \Omega^{\cdot}(P)$ we have then

$$i(x)\omega = \omega\, i(x), \quad x \in \underline{\underline{g}}$$

$$R_g^*\omega = \omega\, \mathrm{Ad}(g^{-1})^*, \quad g \in G$$

i.e. ω is indeed a G-DG-homomorphism by (3.14) (3.16).

It follows immediately that (3.31) induces a DG-homomorphism of the basic subalgebras

$$\omega_H : (\wedge^{\cdot}\underline{\underline{g}}^*)_H \longrightarrow \Omega^{\cdot}(P)_H \tag{3.32}$$

By (3.23) (3.27) this can be written equivalently as

$$\omega_H : \wedge^{\cdot}(\underline{\underline{g}}/\underline{\underline{h}})^{*H} \longrightarrow \Omega^{\cdot}(P/H) . \tag{3.33}$$

Let $\Delta = s^* \circ \omega_H$ be the composition with the map

(3.34) $$s^* : \Omega^\cdot(P/H) \longrightarrow \Omega^\cdot(M)$$

induced by the section s of $P/H \longrightarrow M$ defining the H-reduction of P. Since

(3.35) $$\Delta : \wedge^\cdot(\underline{\underline{g}}/\underline{\underline{h}})^{*H} \longrightarrow \Omega^\cdot(M)$$

is a DG-homomorphism, there is an induced cohomology map

$$\Delta_* : H^\cdot(\underline{\underline{g}},H) \longrightarrow H^\cdot_{DR}(M)$$

which is by definition the generalized characteristic homomorphism of P. □

3.36 COROLLARY. Let the situation be as in theorem 3.30. Then the explicit formula for Δ_* on the cochain level

$$\Delta(\omega) : \wedge^q(\underline{\underline{g}}/\underline{\underline{h}})^{*H} = (\wedge^q\underline{\underline{g}}^*)_H \longrightarrow \Omega^q(M)$$

is given for $\Psi \in (\wedge^q\underline{\underline{g}}^*)_H$ by

(3.37) $$\Delta(\omega)(\Psi) = s^*\Psi(\underbrace{\omega \wedge \ldots \wedge \omega}_{q \text{ factors}})$$

where $\omega \wedge \ldots \wedge \omega$ is the $\wedge^q\underline{\underline{g}}$-valued q-th wedge product of the $\underline{\underline{g}}$-valued 1-form ω with itself.

Δ sends cocycles into cocycles, since it is a DG-homomorphism.

It is clear from the construction that Δ_* depends a priori upon the H-reduction of P given by s. There is one case however where this construction is visibly independent of s. Let $H = K$ be a maximal compact subgroup of G. Then $P/K \longrightarrow M$ has a section $s : M \longrightarrow P/K$, which as a consequence of the contractibility of G/K is unique up to homotopy. We therefore have the following consequence.

3.38 COROLLARY. Let $P \longrightarrow M$ be a flat principal G-bundle and K a maximal compact subgroup of G. Then there is a well-defined multiplicative homomorphism

$$\Delta_* : H^\cdot(\underline{\underline{g}},K) \longrightarrow H^\cdot_{DR}(M) .$$

The same result applies more generally for a closed subgroup $H \dot{\subset} G$ containing a maximal compact subgroup K of G, i.e. $K \subset H \subset G$. Since the fiber H/K of the fibration

$$H/K \longrightarrow P/K \longrightarrow P/H$$

is contractible, the map $P/K \longrightarrow P/H$ is a homotopy equivalence, so that $P/H \longrightarrow M$ has up to homotopy also a unique section. This implies the stated result.

The relative Lie algebra cohomology of the pair (G,H) appears by the construction above as the universal characteristic cohomology of flat G-bundles with an H-reduction. There is a standard complex which realizes this cohomology. All these facts are generalized in the construction of the next chapter, so that we prefer to postpone the discussion of these questions.

The reader wishing to see immediately applications of the constructions of this chapter can turn directly to the end of chapter 4 and to chapter 6. At this point we only mention the geometric significance of the existence of non-trivial classes in the image of Δ_*. It is a measure for the incompatibility of the flat structure on the G-bundle P with the given H-reduction of P. If the flat connection on P is already a G-extension of a flat connection on P', then Δ_* is trivial. This is a consequence of theorem 4.52 applied to this special situation.

4. CHARACTERISTIC CLASSES OF FOLIATED BUNDLES

Let $P \longrightarrow M$ be a foliated principal G-bundle. In this chapter we describe our construction of a characteristic homomorphism [KT 4,6,7] which is a natural generalization of the construction of Δ_* for flat bundles in chapter 3. For ordinary bundles this construction reduces to the ordinary Chern-Weil construction of characteristic classes. For the foliated normal bundle of a foliation this leads to characteristic invariants attached to the foliation.

Let ω be any connection in the principal bundle $P \longrightarrow M$. The multiplicative extension of $\omega : \wedge^1 \underline{\underline{g}}^* \longrightarrow \Omega^1(P)$ leads to an algebra homomorphism $\omega : \wedge^{\cdot} \underline{\underline{g}}^* \longrightarrow \Omega^{\cdot}(P)$. We have seen in proposition 3.5 that this map is a DG-homomorphism precisely if the curvature Ω vanishes. The Weil algebra introduced in [CA] is a DG-algebra $W(\underline{\underline{g}})$ containing $\wedge^{\cdot}\underline{\underline{g}}^*$ and with the property that ω extends to a canonical DG-homomorphism

(4.1) $$k(\omega) : W^{\cdot}(\underline{\underline{g}}) \longrightarrow \Omega^{\cdot}(P),$$

the Weil homomorphism of ω.

The Weil algebra is the tensorproduct

(4.2) $$W(\underline{\underline{g}}) = \wedge \underline{\underline{g}}^* \otimes S\underline{\underline{g}}^*$$

of the exterior algebra $\wedge\underline{\underline{g}}^*$ with the symmetric algebra $S\underline{\underline{g}}^*$. Note that $S^{\cdot}\underline{\underline{g}}^*$ is a graded commutative algebra only if it is viewed as the symmetric algebra over the graded module $\underline{\underline{g}}^*$ with elements of degree two. With this understanding the elements of $S^p\underline{\underline{g}}^*$ are of degree $2p$. The bidegree

$$W^{q,2p}(\underline{\underline{g}}) = \wedge^q \underline{\underline{g}}^* \otimes S^p \underline{\underline{g}}^* \quad \text{for} \quad q,p \geq 0$$

leads to the total degree

$$W^r = \bigoplus_{q+2p=r} W^{q,2p}$$

and $W^{\cdot}(\underline{\underline{g}})$ is then a commutative graded algebra. It is the free commutative algebra over the graded module consisting in degree 1 of the elements $\alpha \in \Lambda^1\underline{\underline{g}}^*$, and in degree 2 of the elements $\tilde{\alpha} \in S^1\underline{\underline{g}}^*$. The canonical isomorphism $\Lambda^1\underline{\underline{g}}^* \longrightarrow S^1\underline{\underline{g}}^*$ is denoted by $\alpha \longrightarrow \tilde{\alpha}$.

4.3 LEMMA [CA]. $W(\underline{\underline{g}})$ _is a G-DG-algebra._

Proof. $\Lambda\underline{\underline{g}}^*$ is already a G-DG-algebra. The operators $i(x)$, $\rho(g)$ on $S\underline{\underline{g}}^*$ are characterized by

$$\begin{aligned} i(x)\tilde{\alpha} &= 0 \\ \rho(g)\tilde{\alpha} &= Ad(g^{-1})^*\tilde{\alpha} \end{aligned} \tag{4.4}$$

for $\tilde{\alpha} \in S^1\underline{\underline{g}}^*$, $x \in \underline{\underline{g}}$, $g \in G$, and the unique extensions of $i(x)$ to the zero derivation and $\rho(g)$ to an automorphism of $S\underline{\underline{g}}^*$.

On the elements $\Lambda^1\underline{\underline{g}}^* \otimes S^1\underline{\underline{g}}^*$ of $W(\underline{\underline{g}})$ let then

$$\begin{aligned} &i(x)(\alpha\otimes 1) = \alpha(x),\ i(x)(1\otimes\tilde{\alpha}) = 0 \\ &\rho(g)(\alpha\otimes 1) = Ad(g^{-1})^*\alpha\otimes 1,\ \rho(g)(1\otimes\tilde{\alpha}) = 1\otimes Ad(g^{-1})^*\tilde{\alpha} \end{aligned} \tag{4.5}$$

The $\underline{\underline{g}}$-representation θ induced by the G-representation ρ is then characterized by being zero on $W^{0,0}$ and by

$$\theta(x)(\alpha\otimes 1) = \theta(x)\alpha\otimes 1,\ \theta(x)(1\otimes\tilde{\alpha}) = 1\otimes\theta(x)\tilde{\alpha}. \tag{4.6}$$

A differential $d = d_W$ of degree 1 is defined on $W(\underline{\underline{g}})$ as the sum

$$d_W = d' + d'' : W^{\cdot} \longrightarrow W^{\cdot+1}$$

where

(4.7) $\quad d' : W^{q,2p} \longrightarrow W^{q+1,2p}, \quad d'' : W^{q,2p} \longrightarrow W^{q-1,2(p+1)}$.

d_W is a derivation, characterized by $d_W | W^{0,0} = 0$ and the formulas

$$d'\alpha = d_\wedge \alpha, \quad i(x)d'\tilde{\alpha} = \theta(x)\tilde{\alpha}$$

(4.8)

$$d''\alpha = \tilde{\alpha}, \quad d''\tilde{\alpha} = 0 \quad .$$

In terms of a basis $x_1, \ldots, x_m \in \underline{g}$ and its dual basis $x_1^*, \ldots, x_m^* \in \underline{g}^*$ the definition of d' reads as follows

(4.9) $\quad (d'\alpha)(x_k, x_e) = -\alpha[x_k, x_e], \quad d'\tilde{\alpha} = \sum_j x_j^* \otimes \theta(x_j)\tilde{\alpha}$.

The first formula is simply the definition of the Chevalley-Eilenberg differential $d = d_\wedge$. The operator $i(x_k)$ applied to the second formula gives

$$i(x_k)d'\tilde{\alpha} = \sum_j i(x_k)x_j^* \otimes \theta(x_j)\tilde{\alpha} = \theta(x_k)\tilde{\alpha}.$$

This implies the second formula in (4.8) and conversely.

Next we show that formula (iv) in the definition (3.12) of a G-DG-algebra holds. It suffices to check it on $\wedge^1\underline{g}^*$ and $S^1\underline{g}^*$. But for $\alpha \in \wedge^1\underline{g}^*$

$$(i(x)d + di(x))\alpha = i(x)d\alpha = i(x)(d_\wedge\alpha + \tilde{\alpha}) = i(x)d_\wedge\alpha = -\alpha[x,-]$$

whereas

$$\theta(x)\alpha = \frac{d}{dt}\Big|_{t=0} \mathrm{Ad}(\exp -tx)^*\alpha = -(\mathrm{ad}\ x)^*\alpha = -\alpha[x,-].$$

For $\tilde{\alpha} \in S^1\underline{g}^*$ we get

$$(i(x)d + di(x))\tilde{\alpha} = \theta(x)\tilde{\alpha} + di(x)\tilde{\alpha} = \theta(x)\tilde{\alpha}.$$

Therefore

(4.10) $$\theta(x) = i(x)d + di(x)$$

holds on $W(\underline{\underline{g}})$. This implies

(4.11) $$\theta(x)d - d\theta(x) = i(x)dd - ddi(x).$$

That this derivation is zero has again only to be checked on $\wedge^1\underline{\underline{g}}^*$ and $S^1\underline{\underline{g}}^*$, which follows easily. Therefore

(4.12) $$\theta(x)d = d\theta(x) = 0$$

It is now easy to verify that $d^2 = 0$ by checking it on $\wedge^1\underline{\underline{g}}^*$ and $S^1\underline{\underline{g}}^*$. For this we use the formula $i(x)dd = ddi(x)$ following from (4.11) (4.12). Finally formulas (ii) (iii) in definition (3.12) are easily verified. □

The Weil algebra satisfies now the following property.

4.13 PROPOSITION [CA]. Let ω be any connection in the principal G-bundle $P \longrightarrow M$. Then there is a G-DG-homomoprhism

(4.14) $$k(\omega) : W^\cdot(\underline{\underline{g}}) \longrightarrow \Omega^\cdot(P)$$

which extends $\omega : \wedge^\cdot \underline{\underline{g}} \longrightarrow \Omega^\cdot(P)$. The map $k(\omega)$ is the Weil homomorphism of ω.

Proof. The curvature

$$\Omega(\omega) = d_P\omega - \omega d_\wedge : \underline{\underline{g}}^* \longrightarrow \Omega^2(P)$$

as defined in (3.4) extends to a unique multiplicative homomorphism $\Omega(\omega) : S^\cdot\underline{\underline{g}}^* \longrightarrow \Omega^{2\cdot}(P)$. Note that $\Omega(\omega)$ preserves total degrees. The Weil homomorphism is then defined on $\wedge\underline{\underline{g}}^* \otimes S\underline{\underline{g}}^*$ by

(4.15) $$k(\omega) = (\omega, \Omega(\omega))$$

It is a homomorphism of graded algebras (preserving total degrees).

It remains to show that with this definition $k(\omega)$ is indeed a G-DG-homomorphism. The compatibility with $i(x)$ and $\theta(x)$ is checked by verifying it separately on $\wedge^1\underline{\underline{g}}^*$ and $S^1\underline{\underline{g}}^*$, which is sufficient. E.g. for $\tilde{\alpha} \in S^1\underline{\underline{g}}^*$ we have $i(x)\tilde{\alpha} = 0$, so that it remains to verify $i(x)\, k(\omega)\tilde{\alpha} = 0$. But

$$i(x)k(\omega)\tilde{\alpha} = i(x)(d_p\omega - \omega d_\wedge)\alpha = (\theta(x) - d_p i(x))\omega(\alpha) - \omega i(x)d_\wedge\alpha$$

$$= \omega\,\theta(x)\alpha - \omega(\theta(x) - d_\wedge i(x))\alpha = d_\wedge i(x)\alpha = 0.$$

The compatibility with d is similarly checked. For $\alpha \in \wedge^1\underline{\underline{g}}^*$ we have

$$k(\omega)d_W\alpha = k(\omega)(d'\alpha + \tilde{\alpha}) = k(\omega)d'\alpha + k(\omega)\tilde{\alpha} = \omega d'\alpha + (d_p\omega - \omega d_\wedge)\alpha = d_p\omega\alpha$$

$$= d_p\, k(\omega)\alpha$$

so that $k(\omega)|\wedge^\cdot\underline{\underline{g}}^*$ commutes with d. For $\tilde{\alpha} \in S^1\underline{\underline{g}}^*$ we have then

$$k(\omega)d_W\tilde{\alpha} = k(\omega)d'\tilde{\alpha} = k(\omega)d'd''\alpha = -k(\omega)d''d'\alpha = -k(\omega)d_W d'\alpha$$

$$= -d_p\,\omega d'\alpha = d_p(d_p\omega - \omega d')\alpha = d_p\, k(\omega)\tilde{\alpha}$$

which shows that $k(\omega)$ is indeed a map of differential algebras. □

In fact the Weil homomorphism $k(\omega) : W^\cdot(\underline{g}) \longrightarrow \Omega^\cdot(P)$ is the unique G-DG-homomorphism extending $\omega : \wedge^\cdot\underline{\underline{g}}^* \longrightarrow \Omega^\cdot(P)$ and the Weil algebra together with the canonical map $\mu : \wedge\underline{\underline{g}}^* \longrightarrow W(\underline{g})$ given by $id \otimes 1$ is the solution to the universal problem characterized by the commutative diagram

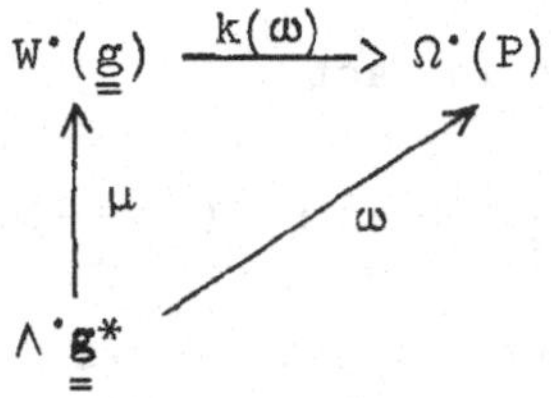

We return to this point of view in chapter 5.

Since $k(\omega)$ is a G-DG-homomorphism, it induces a homomorphism on G-basic elements

$$k(\omega)_G : W^{\cdot}(\underline{\underline{g}})_G \longrightarrow \Omega^{\cdot}(P)_G$$

As $S\underline{\underline{g}}^*$ is precisely the set of elements in $W(\underline{\underline{g}})$ killed by $i(x)$ for $x \in \underline{\underline{g}}$, it follows that

$$W^{\cdot}(\underline{\underline{g}})_G = (S^{\cdot}\underline{\underline{g}}^*)^G .$$

The elements in $(S^p\underline{\underline{g}}^*)^G$ are symmetric p-linear forms Φ on $\underline{\underline{g}}$, which are invariant under the G-action induced by the adjoint representation, i.e.

$$\Phi(\mathrm{Ad}(g)x_1,\ldots,\mathrm{Ad}(g)x_p) = \Phi(x_1,\ldots,x_p)$$

for $x_1,\ldots,x_p \in \underline{\underline{g}}$ and $g \in G$. Infinitesimally this invariance condition reads

$$\sum_{j=1}^{p} \Phi(x_1,\ldots,[x,x_j],\ldots,x_p) = 0$$

for $x_1,\ldots,x_p \in \underline{\underline{g}}$ and $x \in \underline{\underline{g}}$. The algebra

$$I(G) = (S\underline{\underline{g}}^*)^G$$

is the algebra of invariant polynomials. Note that with the natural degree convention in $S^{\cdot}\underline{\underline{g}}^*$ we have $I^{2p}(G) = (S^p\underline{\underline{g}}^*)^G$.

On the target complex we have on the other hand $\Omega^{\cdot}(P)_G = \Omega^{\cdot}(M)$, so that in fact

(4.16) $$k(\omega)_G \equiv h(\omega) : I^{\cdot}(G) \longrightarrow \Omega^{\cdot}(M).$$

The formula $\theta(x) = di(x) + i(x)d$ shows first that $d : I(G) \to I(G)$. But for $\Phi \in W^{0,2p}$ clearly $d''\Phi = 0$, so that $d\Phi = d'\Phi \in W^{1,2p}$ and hence $d\Phi = 0$. Thus $d|I(G) = 0$. It follows that $h(\omega)$ induces a cohomology map

$$h_* : I^{\cdot}(G) \longrightarrow H^{\cdot}_{DR}(M).$$

$h(\omega)$ resp. h_* is the Chern-Weil homomorphism of P ([C1] [C2]). We repeat its definition. On the cochain level $h(\omega)$ is the map induced on G-basic elements by the Weil homomorphism $k(\omega)$:

(4.17) $$\begin{array}{ccc} W^{\cdot}(\underline{\underline{g}}) & \xrightarrow{k(\omega)} & \Omega^{\cdot}(P) \\ \cup & & \cup \\ I^{\cdot}(G) & \xrightarrow{h(\omega)} & \Omega^{\cdot}(M) \end{array}$$

$k(\omega)$ induces of course also a cohomology map. But $H(W(\underline{\underline{g}})) \cong \mathbb{R}$, so that this map is trivial.

The explicit definition of $h(\omega)$ is given by

(4.18) $$h(\omega)\Phi = \Phi(\underbrace{\Omega \wedge \ldots \wedge \Omega}_{p \text{ factors}})$$

for $\Phi \in I^{2p}(G)$, where $\Omega \wedge \ldots \wedge \Omega$ is the $S^p\underline{\underline{g}}$-valued p-th exterior product with itself of the $S^1\underline{\underline{g}}$-valued curvature 2-form Ω. For a flat connection ω the map $h(\omega)$ is visibly trivial.

The analogue of this construction on the space level is given by the classifying map of the G-bundle $P \to M$ into the universal G-bundle $E_G \longrightarrow B_G$ over the classfying space B_G of G:

$$\begin{array}{ccc} P & \longrightarrow & E_G \\ \downarrow & & \downarrow \\ M & \longrightarrow & B_G \end{array}$$

This diagram induces in DeRham cohomology the commutative diagram

$$(4.19) \qquad \begin{array}{ccc} \Omega^{\cdot}(E_G) & \longrightarrow & \Omega^{\cdot}(P) \\ \uparrow & & \uparrow \\ \Omega^{\cdot}(B_G) & \longrightarrow & \Omega^{\cdot}(M) \end{array}$$

where the left hand side has of course to be taken with a grain of salt in view of the infinite dimensionality of the spaces E_G and B_G. (4.17) is the algebraic analogue of (4.19), i.e. the Weil algebra $W(\underline{\underline{g}})$ plays the role of the DeRham complex of the total space E_G of a universal G-bundle. The contractibility of E_G is reflected by the property $H(W(\underline{\underline{g}})) \cong \mathbb{R}$. Indeed, ignoring questions of infinite dimensionality, a connection in the universal bundle $E_G \longrightarrow B_G$ induces by (4.17) a commutative diagram

$$(4.20) \qquad \begin{array}{ccc} W^{\cdot}(\underline{\underline{g}}) & \longrightarrow & \Omega^{\cdot}(E_G) \\ \cup & & \cup \\ I^{\cdot}(G) & \longrightarrow & \Omega^{\cdot}(B_G) \end{array}$$

The bottom map is known to be an isomorphism in cohomology for a compact connected Lie group G [C2].

All these facts make sense for any connection ω in an ordinary bundle P. Assume now that $P \longrightarrow M$ is a foliated bundle and ω an adapted connection. To explain the new feature appearing, we need the following definitions [KT 4,6,7].

$W(\underline{\underline{g}})$ has an even decreasing filtration by ideals

(4.21) $$F^{2p}W(\underline{\underline{g}}) = S^p\underline{\underline{g}}^* . W(\underline{\underline{g}}), \quad p \geq 0$$

i.e. $$F^{2p}W(\underline{\underline{g}}) = \bigoplus_{j \geq p} \wedge^{\cdot}\underline{\underline{g}}^* \otimes S^j\underline{\underline{g}}^*$$

Note that $F^{2p}W$ is the p-th power of the ideal F^2W and that the ideals $F^{2p}W$ are closed under d_W, i.e. differential ideals. These ideals are in fact G-DG-ideals.

$\Omega^{\cdot}(P)$ has a decreasing filtration by ideals defined by the sheaf $\underline{Q}^*$ of 1-forms annihilating the foliation on the base space M of the bundle $\pi : P \longrightarrow M$. It is given by

(4.22) $$F^p\Omega^{\cdot}(P) = \Gamma(P, \pi^*\wedge^p\underline{Q}^* . \Omega^{\cdot}_P), \quad p \geq 0.$$

Clearly $F^p\ \Omega^{\cdot}(P)$ is the p-th power of the ideal $F^1\ \Omega^{\cdot}(P)$ and the ideals $F^p\ \Omega^{\cdot}(P)$ are closed under d_P. They are in fact G-DG-ideals.

4.23 THEOREM [KT 4,6,7]. *Let* $P \longrightarrow M$ *be a foliated principal G-bundle and* ω *an adapted connection on* P. *Then the Weil-homomorphism* $k(\omega) : W^{\cdot}(\underline{\underline{g}}) \longrightarrow \Omega^{\cdot}(P)$ *is filtration-preserving in the sense that*

(4.24) $$k(\omega)\ F^{2p}W^{\cdot}(\underline{\underline{g}}) \subset F^p\ \Omega^{\cdot}(P), \quad p \geq 0.$$

If ω *is moreover a basic connection, then*

(4.25) $$k(\omega)\ F^{2p}\ W^{\cdot}(\underline{\underline{g}}) \subset F^{2p}\ \Omega^{\cdot}(P), \quad p \geq 0.$$

Proof. Since the filtrations and $k(\omega)$ are multiplicative, and $(F^{2p}W) = (F^2W)^p$, it suffices to verify this property for $p = 1$. But

$$F^2W(\underline{\underline{g}}) = \wedge^{\cdot}\underline{\underline{g}}^* \otimes S^+\underline{\underline{g}}^*$$

so it suffices to check this property on $S^1\underline{\underline{g}}^*$.

For $\tilde{\alpha} \in S^1\underline{\underline{g}}^*$ we have

$$k(\omega)\tilde{\alpha} = \Omega(\omega)\alpha = \alpha\Omega.$$

But for an adapted connection we have shown in (1.42) that

$$\alpha\Omega \in F^1\Omega^2(P).$$

If ω is moreover basic, we have shown in (2.27) that

$$\alpha\Omega \in F^2\Omega^2(P)$$

This finishes the proof. □

If q denotes the codimension of the foliation on M, it is clear from (4.22) that

(4.26) $$F^{q+1}\Omega^{\cdot}(P) = 0.$$

4.27 COROLLARY. <u>For an adapted connection</u> ω

$$k(\omega)F^{2(q+1)}W^{\cdot}(\underline{\underline{g}}) = 0.$$

<u>For a basic connection</u> ω <u>moreover</u>

$$k(\omega)F^{2([q/2]+1)}W^{\cdot}(\underline{\underline{g}}) = 0.$$

Since both filtrations and $k(\omega)$ preserve the G-DG-structure, the same facts hold for the induced filtration on G-basic elements. We note that

(4.28) $$F^{2p}I^{\cdot}(G) = \bigoplus_{j \geq p} I^{2j}(G)$$

(4.29) $$F^p\Omega^\cdot(M) = \Gamma(M, \wedge^p \underline{Q}^* \cdot \Omega^\cdot_M).$$

We have then the following result.

4.30 COROLLARY. Let $P \longrightarrow M$ be a foliated principal G-bundle and ω an adapted connection on P. Then the Chern-Weil homomorphism $h(\omega) : I^\cdot(G) \longrightarrow \Omega^\cdot(M)$ is filtration-preserving in the sense that

(4.31) $$h(\omega)F^{2p}I^\cdot(G) \subset F^p\Omega^\cdot(M), \quad p \geqq 0.$$

If ω is moreover a basic onnection, then

(4.32) $$h(\omega)F^{2p}I^\cdot(G) \subset F^{2p}\Omega^\cdot(M), \quad p \geqq 0.$$

If q denotes again the codimension of the foliation on M, it is clear from (4.29) that

(4.33) $$F^{q+1}\Omega^\cdot(M) = 0.$$

Therefore in particular

(4.34) $$h(\omega)F^{2(q+1)}I^\cdot(G) = 0 \text{ for an adapted connection } \omega.$$

(4.35) $$h(\omega)F^{2([q/2]+1)}I^\cdot(G) = 0 \text{ for a basic connection } \omega.$$

This vanishing phenomenon was discovered by Bott for the normal bundle of a foliation [B1] [B2].

4.36 COROLLARY (Bott's Vanishing Theorem). Let Q be the normal bundle of a codimension q-foliation on M. Then for the characteristic ring $\mathrm{Pont}^\cdot(Q) = h_*(I^\cdot(O(q))) \subset H^\cdot_{DR}(M)$ the following holds:

$$\mathrm{Pont}^p(Q) = 0 \quad \text{for} \quad p > 2q.$$

This follows from the existence of the Bott connection (2.15) in Q, which canonically foliates Q. The Pontrjagin classes of Q are the characteristic classes of the orthogonal frame bundle of Q.

If the Bott connection on Q extends to a basic connection, then by (4.35)

$$\mathrm{Pont}^p(Q) = 0 \quad \text{for} \quad p > q.$$

This improvement on Bott's vanishing theorem was observed by Molino [MO 1]. As explained briefly in 2.43, the existence of a basic connection in Q is characterized by the vanishing of a cohomology class $\zeta \in H^1(M, \underline{L}; \underline{\mathrm{Hom}}_{\underline{O}}(\underline{P}(\underline{\underline{g}}^*), \underline{Q}^*))$ [KT 3]. The first example of such a situation has been given by Pasternack [P1] [P2] for Riemannian foliations (see 2.35). The normal bundle Q inherits then a basic connection from the Riemannian connection on the target manifold of the local submersions defining the foliation (see [P2, §4]).

Bott further proved the following result [B1] [B2].

4.37 Bott's Vanishing Theorem in the complex case. *Let M be a complex manifold with a complex foliation of complex codimension q. Then for the characteristic ring $\mathrm{Chern}^{\cdot}(Q) = h_*(I^{\cdot}(GL(g,\mathbb{C}))) \subset H^{\cdot}_{DR}(M)$ of the normal bundle Q the following holds:*

$$\mathrm{Chern}^p(Q) = 0 \quad \text{for} \quad p > 2q.$$

This result is a consequence of the filtration preserving property proposition 3.4 in [KT 7] generalizing theorem 4.23 above. The original proof in [B1] proceeds by extending the Bott connection in Q to a smooth connection of type (1,0). The result follows then from the decomposition of the curvature into a form of type (2,0) and a form of type (1,1).

Bott used these results to answer the following question posed by Haefliger [H2]. Is any subbundle $E \subset T_M$ isomorphic to an involutive subbundle $L \subset T_M$? The vanishing theorem shows that the involutivity of L imposes conditions on the characteristic ring of $Q = T_M/L$. But isomorphic subbundles $E \cong L$ have homotopic classifying maps. It follows that the necessary conditions imposed by Bott's result must also hold for a subbundle $E \subset T_M$ which is isomorphic to an involutive one. Bott gave examples of subbundles not satisfying those characteristic class conditions, hence not isomorphic to involutive subbundles (see [B1] [B2]). The vanishing theorem was an important step in the development of the theory. In fact one of the points of our construction in [KT 4,6,7] is that this vanishing phenomenon can be interpreted as a consequence of the filtration-preserving property 4.23 of the Weil homomorphism, and that this leads directly to the existence of new characteristic classes, as we now proceed to explain in detail.

We define for every $k \geq 0$ the truncated Weil algebra

$$(4.38) \qquad W(\underline{\underline{g}})_k = W(\underline{\underline{g}})/F^{2(k+1)}\, W(\underline{\underline{g}}).$$

From Theorem 4.23 we obtain first the following consequence [KT 4,6, 7].

4.39 COROLLARY. <u>Let</u> $P \longrightarrow M$ <u>be a foliated bundle and</u> q <u>the codimension of the foliation of</u> M. <u>Then an adapted connection</u> ω <u>defines a</u> G-DG-<u>homomorphism</u> $k(\omega) : W^\cdot(\underline{\underline{g}})_q \longrightarrow \Omega^\cdot(P)$ <u>which induces a cohomology map</u>

$$k_* : H^\cdot(W(\underline{\underline{g}})_q) \longrightarrow H^\cdot_{DR}(P)$$

<u>If</u> ω <u>is moreover a basic connection, then</u> $k(\omega) : W^\cdot(\underline{\underline{g}})_{[q/2]} \rightarrow \Omega^\cdot(P)$ <u>and</u>

$$k_* : H^{\cdot}(W(\underline{\underline{g}})_{[q/2]}) \longrightarrow H^{\cdot}_{DR}(P).$$

Whereas the cohomology of $W(\underline{\underline{g}})$ is trivial, this is no longer so for the cohomology of the truncated algebra $W(\underline{\underline{g}})_q$. The map k_* furnishes cohomology classes on P which are defined on the cochain level by

$$\Psi(\underbrace{\omega \wedge \ldots \wedge \omega}_{s \text{ factors}} \wedge \underbrace{\Omega \wedge \ldots \wedge \Omega}_{r \text{ factors}})$$

for $\Psi \in W^{s,r}(\underline{\underline{g}}) = \wedge^s \underline{\underline{g}}^* \otimes S^r \underline{\underline{g}}^*$. Here the exterior product $\omega \wedge \ldots \wedge \omega \wedge \Omega \wedge \ldots \wedge \Omega$ is taken with respect to the multiplication in $\wedge^s \underline{\underline{g}} \otimes S^r \underline{\underline{g}}$.

For any subgroup $H \subset G$ we define now the relative Weil algebra

(4.40) $$W(\underline{\underline{g}},H) = W(\underline{\underline{g}})_H.$$

Since the canonical filtration (4.21) of W is by G-DG-ideals, it induces a canonical filtration

(4.41) $$F^{2p}W(\underline{\underline{g}},H) = F^{2p}W(\underline{\underline{g}}))_H.$$

For every $k \geq 0$ we have then the truncated relative Weil algebra

(4.42) $$W(\underline{\underline{g}},H)_k = W(\underline{\underline{g}},H)/F^{2(k+1)}W(\underline{\underline{g}},H) \cong (W(\underline{\underline{g}})_k)_H.$$

We have then the following natural generalization of theorem 3.30.

4.43 THEOREM [KT 6,7]. <u>Let</u> $\pi : P \longrightarrow M$ <u>be a foliated principal</u> G-<u>bundle</u>, $H \subset G$ <u>a closed subgroup and</u> P' <u>an</u> H-<u>reduction of</u> P <u>given by a section</u> $s : M \longrightarrow P/H$ <u>of the induced map</u> $\hat{\pi} : P/H \longrightarrow M$.

(i) <u>There is a well-defined multiplicative homomorphism</u>

$$\Delta_* : H^\cdot(W(\underline{\underline{g}},H)_q) \longrightarrow H^\cdot_{DR}(M)$$

<u>where</u> q <u>is the codimension of the foliation on</u> M. Δ_* <u>is the generalized characteristic homomorphism of</u> P.

(ii) Δ_* <u>does not depend on the choice of an adapted connection in</u> P. <u>But if</u> P <u>admits a basic connection, then</u>

$$\Delta_* : H^\cdot(W(\underline{\underline{g}},H)_{[q/2]}) \longrightarrow H^\cdot_{DR}(M).$$

(iii) Δ_* <u>is functorial under pullbacks and functorial in</u> (G,H).

(iv) Δ_* <u>is invariant under integrable homotopies</u>.

Proof. The construction of Δ_* should by now be clear. For any adapted connection ω in P we have a DG-homomorphism

$$k(\omega)_H : W^\cdot(\underline{\underline{g}},H) \longrightarrow \Omega^\cdot(P)_H = \Omega^\cdot(P/H)$$

which is still filtration preserving. The induced map

$$k(\omega)_H : W^\cdot(\underline{\underline{g}},H)_q \longrightarrow \Omega^\cdot(P/H)$$

composes with

$$s^* : \Omega^\cdot(P/H) \longrightarrow \Omega^\cdot(M)$$

to a DG-homomorphism

(4.44) $$\Delta(\omega) = s^* \circ k(\omega)_H : W^\cdot(\underline{\underline{g}},H)_q \longrightarrow \Omega^\cdot(M)$$

which on the cohomology level defines Δ_*.

The independence of Δ_* from the choice of ω will be discussed later. It follows from a universal homotopy construction which in particular gives a homotopy through adapted connections between any two adapted connections. The existence of a basic connection ω clearly implies a factorization of $\Lambda(\omega)$ as in the diagram

$$\begin{array}{ccc} W(\underline{\underline{g}},H)_q & \xrightarrow{\Delta(\omega)} & \Omega^{\cdot}(M) \\ \downarrow & \nearrow & \\ W(\underline{\underline{g}},H)_{[q/2]} & & \end{array}$$

The functoriality under pullbacks means more precisely the following. Let $P \longrightarrow M$ be a foliated bundle and $f : M' \longrightarrow M$ a map transversal to the foliation of M. The pullback bundle $P' = f^*P \longrightarrow M'$ is then canonically foliated with respect to the pullback foliation (see 2.23). The pullback foliation on M' has the same codimension q as the original foliation on M. The H-reduction of P defined by the section s of $P/H \longrightarrow M$ defines an H-reduction of f^*P via the section f^*s of $f^*P/H \longrightarrow M'$. Then there is a commutative diagram

$$\begin{array}{ccc} & \xrightarrow{\Delta_*(P)} & H^{\cdot}_{DR}(M) \\ H(W(\underline{\underline{g}},H)_q) & & \downarrow f^* \\ & \xrightarrow{\Delta_*(f^*P)} & H^{\cdot}_{DR}(M') \end{array}$$

The contravariant functoriality under maps $(G,H) \longrightarrow (G',H')$ is to be understood in the obvious sense. It is discussed in detail in 4.59 and the paragraphs following.

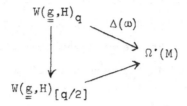

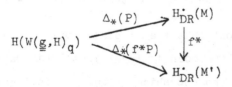

For the last statement we need to explain the concept of an integrable homotopy of foliated bundles. Let $P_i \longrightarrow M$ $(i = 0,1)$ be G-bundles foliated with respect to codimension q foliations on M. The foliated bundles P_0, P_1 are integrably homotopic, if there exists a foliated bundle $P \longrightarrow M \times [0,1]$ over a codimension q foliation of $M \times [0,1]$, such that with the canonical maps $j_\tau : M \longrightarrow M \times [0,1]$ given by $j_\tau(x) = (x,\tau)$ we have $P_i \cong j_i^*P$ $(i = 0,1)$. The maps j_i are further supposed to be transversal to the foliation on $M \times [0,1]$. P is in addition supposed to be equipped with an H-reduction, pulling back to H-reductions of P_i for $i = 0,1$. The generalized characteristic homomorphism $\Delta_*(P_i)$ for $i = 0,1$ is then the composition

$$\Delta_*(Pi) = j_i^* \, \Delta_*(P) .$$

But the composition

$$\begin{array}{ccc} H(W(\underline{\underline{g}},H)_q) & \xrightarrow{\Delta_*(P)} & H_{DR}(M \times [0,1]) \\ & \searrow & \downarrow {\scriptstyle j_\tau{}^*} \\ & & H_{DR}(M) \end{array}$$

is independent of τ, since j_τ^* is independent of τ. Therefore $\Delta_*(P_0) = \Delta_*(P_1)$. This finishes the proof of theorem 4.43. □

It is worth pointing out that the crucial property needed in the construction of Δ_* on the cochain level is the filtration preserving property of the Weil homomorphism $k(\omega)$. In fact any connection ω and any filtration on the DeRham complex of P with this property will do. This remark is useful in certain applications, where these properties can be established for connections which are not verifiably adapted. An example of this is Martinet's construction in [MR] which works for precisely this reason.

4.45 COROLLARY. <u>Let the situation be as in theorem 4.43. Then the evaluation of the generalized characteristic homomorphism</u>

$$\Delta(\omega) : W^{s,2r}(\underline{\underline{g}},H) \longrightarrow \Omega^{s+2r}(M)$$

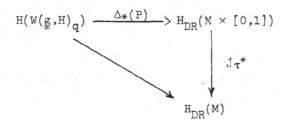

on the cochain level is given for $\Psi \in \wedge^s \underline{\underline{g}}^* \otimes S^r \underline{\underline{g}}^*$ *by the formula*

(4.46) $$\Delta(\omega)(\Psi) = s^*\Psi(\underbrace{\omega \wedge \cdots \wedge \omega}_{s \text{ factors}} \wedge \underbrace{\Omega \wedge \cdots \wedge \Omega}_{r \text{ factors}}).$$

For a flat bundle P the codimension $q = 0$ and

$$F^2 W(\underline{\underline{g}}) = S^+\underline{\underline{g}}^* . W(\underline{\underline{g}}).$$

Therefore $W(\underline{\underline{g}})_0 \cong \wedge \underline{\underline{g}}^*$ and more generally

(4.47) $$W(\underline{\underline{g}},H)_0 \cong (\wedge \underline{\underline{g}}^*)_H = \wedge(\underline{\underline{g}}/\underline{\underline{h}})^{*H}.$$

It is clear that in this case our construction of Δ_* reduces to the constructions of chapter 3, after which they have been patterned.

As in chapter 3 for a closed subgroup $H \subset G$ containing a maximal compact subgroup K of G the generalized characteristic homomorphism Δ_* depends only on P (see corollary 3.38).

4.48 CHARACTERISTIC CLASSES OF FOLIATIONS. This construction of Δ_* applies in particular to the foliated frame bundle $F(Q)$ of the normal bundle Q of a foliation. For $q = \dim Q$ there is then a homomorphism

$$\Delta_* : H^\cdot(W(\underline{\underline{gl}}(q),O(q))_q) \longrightarrow H^\cdot_{DR}(M)$$

giving invariants attached to the foliation.

The first non-trivial example of such a characteristic class for a foliation was given by Godbillon-Vey in [GV] (see chapter 7, section 7.7). They also pointed out relations with the cohomology $H(\mathfrak{L}_0(\mathbb{R}^q))$ of the Lie algebra $\mathfrak{L}_0(\mathbb{R}^q)$ of formal vectorfields on $\mathbb{R}^q$ computed by Gelfand-Fuks [GF 2]. This construction was generalized by Bott-Haefliger [BH] [H 5] to the construction of a generalized characteristic homomorphism for a foliation, defined on the Gelfand-Fuks cohomology $H(\mathcal{L}_0(\mathbb{R}^q),O(q))$. Independent construction of characteristic classes for foliations have also been given by Bernstein-Rosenfeld [BR 1] [BR 2] and Malgrange (not published).

The simplest way to relate the construction of Bott-Haefliger to the construction presented here is to think of the Gelfand-Fuks complex $\Lambda\mathcal{L}_0(\mathbb{R}^q)^*$ as possessing the formal properties of the DeRham complex of a foliated principal bundle with structural group $GL(q)$. There is then a canonical homomorphism

$$(4.49) \qquad H(W(\underline{\underline{gl}}(q),O(q))_q) \longrightarrow H(\mathcal{L}_0(\mathbb{R}^q),O(q))$$

which in fact is an isomorphism by a result of Gelfand-Fuks [GF 2]. See also Guillemin [GN] and Losik [LK]. The structure of $H(W(\underline{\underline{gl}}(q),O(q))_q)$ has been determined by Vey in [GB 2], whereas the authors have determined the structure of $H(W(\underline{\underline{g}},H)_q)$ for a wide class of reductive pairs and arbitrary truncation index q in [KT 5] (see chapter 5). The isomorphism above identifies the two constructions of characteristic homomorphisms of foliations by Godbillon-Vey-Bott-Haefliger on $H(\mathcal{L}_0(\mathbb{R}^q),O(q))$ and by the authors on $H(W(\underline{\underline{gl}}(q),O(q))_q)$. Both constructions have their advantages. The definition of a characteristic homomorphism on $H(\mathcal{L}_0(\mathbb{R}^q),O(q))$ is adapted to the definition of a foliation as a Γ_q-cocycle (see 1.21) in the spirit of Haefliger and can be varied accordingly (see [H 5]). The definition of ${}^{\prime}\Delta_*$ as presented here is on the other hand completely functorial in (G,H) and suits itself in view of this flexibility extraordinarily well for a wide variety of geometric applications.

As an illustration consider the normal bundle Q of a Riemannian foliation on M (see 2.35). By the geometric interpretation in theorem 4.52 below, the generalized characteristic homomorphism

$$\Delta_*(Q) : H(W(\underline{\underline{gl}}(q),O(q))_{[q/2]}) \longrightarrow H_{DR}(M)$$

is induced from the ordinary characteristic homomorphism

$$h_*(Q) : I(O(g))_{[q/2]} \longrightarrow H_{DR}(M)$$

and thus Δ_* is trivial on the ideal $H^{\cdot}(K_q)$ of secondary invariants (see 4.52, (ii)).

But the orthogonal frame bundle of Q is also foliated and in fact equipped with a basic connection, so that there is by our construction also a generalized characteristic homomorphism

$$\Delta_*(Q) : H(W(\underline{\underline{so}}(q),H)_{[q/2]}) \longrightarrow H_{DR}(M)$$

for any reduction of the $O(q)$-frame bundle of Q to a subgroup $H \subset O(q)$. An example is a Riemannian foliation with a trivial normal bundle, where $H = \{e\}$. This map $\Delta_*(Q)$ gives rise to highly non-trivial secondary characteristic invariants, as shown in [KT 9,10] (see chapter 7 of these notes).

4.50 DERIVED CHARACTERISTIC CLASSES. The authors would like to point out that the construction presented here gives in fact much more then just a definition of Δ_*. Since on the cochain level the characteristic homomorphism is filtration preserving, it induces a map of the spectral sequences associated to the corresponding filtered complexes. Thus the generalized characteristic homomorphism is just the map in total cohomology, whereas there are also induced maps on all levels of the spectral sequences. These are the derived characteristic homomorphisms

$$\Delta_* : E_{2r}^{2p,n-2p}(W(\underline{\underline{g}},H)_q) \longrightarrow E_r^{p,n-p}(\Omega(M))$$

defined in [KT 7], (6.1) for $r \geq 1$. In particular the derived characteristic homomorphism Δ_1 on the basis terms $E_2^{2p,0}(W) \cong I^{2p}(G)_q$ of the initial term

$$E_2^{2s,t}(W) \cong I^{2s}(G)_q \otimes H^t(\underline{\underline{g}},H)$$

of the spectral sequence of $W(\underline{\underline{g}},H)_q$ is intimately related to the obstruction class $\zeta \in H^1(M,\underline{L};\underline{Hom_0}(\underline{P}(\underline{\underline{g}}^*),\underline{Q}^*))$ for basic connections described in 2.43. This observation was made in the note [KT 4]. More details can be found in [KT 7], section 7.

The spectral sequence defined by the filtration (4.29) of $\Omega^\cdot(M)$ is a generalization of the Leray spectral sequence which one obtains if the quotient map $M \longrightarrow M/\mathcal{L}$ onto the space of leaves is a fibration. This is made precise (in a slightly more general context) in proposition 5.17 of [KT 7]. See section 5 of that paper for a detailed discussion of this spectral sequence and references to the work of Reinhart [RE 1], Molino [MO 1] and Vaisman [VZ 1,2].

4.51 THE GEOMETRIC SIGNIFICANCE OF Δ_*. We discuss next the geometric significance of the existence of non-trivial elements in the image of Δ_*. For this purpose we need to assume that the pair $(\underline{\underline{g}},\underline{\underline{h}})$ associated to (G,H) is a reductive pair of Lie algebras. This means that $\underline{\underline{g}}$ is a reductive Lie algebra and $\underline{\underline{h}}$ reductive in $\underline{\underline{g}}$, i.e. the adjoint representation of $\underline{\underline{h}}$ in $\underline{\underline{g}}$ is semisimple. $\underline{\underline{h}}$ is then itself a reductive Lie algebra. We have then the following result.

4.52 THEOREM [KT 6,7]. Let P be a foliated G-bundle, $H \subset G$ a closed subgroup with finitely many connected components and P' an H-reduction of P. Assume $(\underline{\underline{g}},\underline{\underline{h}})$ to be a reductive pair of Lie algebras.

(i) Then there is a split exact sequence of algebras

$$(4.53) \qquad 0 \longrightarrow H(K_q) \longrightarrow H(W(\underline{\underline{g}},H)_q) \underset{\kappa}{\overset{\longleftarrow}{\longrightarrow}} I(G)_q \otimes_{I(G)} I(H) \longrightarrow 0$$

and the composition $\Delta_*(P) \circ \kappa$ is induced by the characteristic homomorphism

$$h_*(P') : I(H) \longrightarrow H_{DR}(M) \quad \text{of} \quad P'.$$

(ii) If the foliation of the G-bundle P is induced by a foliation of the H-reduction P', then

$$\Delta_* | H(K_q) = 0.$$

$H(K_q)$ is the algebra of secondary characteristic invariants.

Thus the non-triviality of $\Delta_* | H(K_q)$ is a measure for the incompatibility of the two geometric structures given by (a) the foliated structure of P and (b) the H-reduction of P. If they are compatible, i.e. the foliation of P is obtained from a foliation of P' by extension of the structural group, then Δ_* gives nothing more than the ordinary Chern-Weil homomorphism of the H-reduction P'. This geometric idea is underlying all applications of our construction, as it is amply demonstrated in later chapters. The precise meaning of the complex K_q is explained in the course of the following proof, which relies on the results of chapter 5. For computational purposes it is an essential feature that the construction of Δ_* on the cochain level by an adapted connection ω gives an explicit realization of $\Delta(\omega) | K_q$ on the cochain level.

Proof of theorem 4.52. The proof of (ii) is based on the functoriality of Δ_* in (G,H) (see 4.59 for more details). For the map $(H,H) \longrightarrow (G,H)$ there is an induced map

$$W(\underline{\underline{g}},H)_q \longrightarrow W(\underline{\underline{h}},H)_q = I(H)_q$$

A factorization of Δ_* as in the diagram

$$(4.54)\qquad \begin{array}{ccc} H(W(\underline{\underline{g}},H)_q) & \xrightarrow{\Delta_*} & H_{DR}(M) \\ \downarrow & \nearrow & \\ I(H)_q & & \end{array}$$

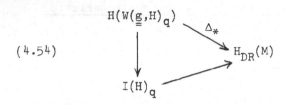

will take place if the foliation in the G-bundle P is already induced from a foliation in the H-reduction P'. Then Δ_* vanishes on the kernel of the vertical homomorphism.

The precise statement of theorem 4.52 is a refinement of this argument. It uses the results of chapter 5 on the computations of $H(W(\underline{\underline{g}},H)_q)$, in particular the complex

$$A = \wedge P \otimes I(G)_q \otimes I(H)$$

with the differential d_A realizing the cohomology of $W(\underline{\underline{g}},H)_q$ by theorem 5.85. The inclusion $i : \underline{\underline{h}} \subset \underline{\underline{g}}$ induces a restriction homomorphism $W(i) : W(\underline{\underline{g}}) \longrightarrow W(\underline{\underline{h}})$, and further $W(i) : W(\underline{\underline{g}},H)_q \longrightarrow I(H)_q$, which appears in the following commutative diagram

$$(4.55)\qquad \begin{array}{ccc} W(\underline{\underline{g}},H)_q & \xrightarrow{W(i)} & I(H)_q \\ \uparrow \varphi & & \uparrow \\ A & \xrightarrow{p} & I(G)_q \otimes_{I(G)} I(H) \end{array}$$

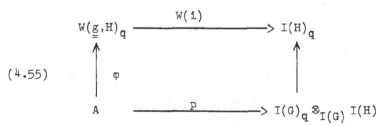

The map $\varphi : A \longrightarrow W(\underline{\underline{g}},H)_q$ denotes the homology equivalence (5.82) and p is induced by the canonical projection $A \to I(G)_q \otimes I(H)$ along $\wedge P$. The vertical map on the right is the canonical projection of

$$(4.56) \qquad I(G)_q \otimes_{I(G)} I(H) \cong I(H)/F^{2(q+1)}\, I(G)\,.\,I(H)$$

onto $I(H)_q = I(H)/F^{2(q+1)}\, I(H)$. Here $I(G)\,.\,I(H)$ denotes the $I(G)$-module structure on $I(H)$ induced by the restriction $i^* : I(G) \longrightarrow I(H)$. Diagram (4.55) gives rise to the factorization

$$H(W(\underline{\underline{g}},H)_q) \longrightarrow I(G)_q \otimes_{I(G)} I(H) \longrightarrow I(H)_q$$

of the vertical homomorphism in (4.54). With the definition

$$(4.57) \qquad K_q = \ker(A \longrightarrow I(G)_q \otimes_{I(G)} I(H))$$

it follows then that the sequence

$$0 \longrightarrow H(K_q) \longrightarrow H(W(\underline{\underline{g}},H)_q) \longrightarrow I(G)_q \otimes_{I(G)} I(H) \longrightarrow 0$$

is exact. We note in passing that all classes in $H(K_q)$ are already realized by cocycles in the subalgebra (not a subcomplex)

$$\wedge^+ P \otimes I(G)_q \otimes I(H) = \ker(A \to I(G)_q \otimes I(H)).$$

For this see the comments in chapter 5 following the discussion of the difference construction for Δ_*. For computational purposes it is of course important to have such a subalgebra giving already rise to all secondary characteristic invariants.

Next we discuss the splitting of the exact sequence (4.53). By averaging there exists an H-equivariant splitting $\theta : \underline{\underline{g}} \longrightarrow \underline{\underline{h}}$ of the exact sequence $0 \longrightarrow \underline{\underline{h}} \longrightarrow \underline{\underline{g}} \longrightarrow \underline{\underline{g}}/\underline{\underline{h}} \longrightarrow 0$ of H-modules. It induces an H-DG-homomorphism

$$k(\theta) : W(\underline{\underline{h}}) \longrightarrow W(\underline{\underline{g}}),$$

the Weil-homomorphism of the (formal) connection

$u \circ \wedge\theta^* : \wedge\mathfrak{h}^* \longrightarrow \wedge\mathfrak{g}^* \longrightarrow W(\mathfrak{g})$, where $u : \wedge\mathfrak{g}^* \longrightarrow W(\mathfrak{g})$ denotes the canonical inclusion (see section 5.11 in chapter 5 for the concept of a formal connection). On H-basic elements $k(\theta)$ induces $k(\theta)_H : I(H) \longrightarrow W(\mathfrak{g},H)$. Then there is an induced map

$$(\mathrm{can},\ k(\theta)_H) : I(G)_q \otimes I(H) \longrightarrow W(\mathfrak{g},H)_q .$$

Consider the diagram

$$(4.58)\qquad \begin{array}{ccc} W(\mathfrak{g},H)_q & \xleftarrow{\ (\mathrm{can},\ k(\theta)_H)\ } & I(G)_q \otimes I(H) \\ \cong\ \uparrow \varphi & \overset{\kappa}{\swarrow} & \downarrow \mathrm{proj} \\ A & \xrightarrow{\quad p \quad} & I(G)_q \otimes_{I(G)} I(H) \end{array}$$

where the maps φ and p are as in (4.55), and the right vertical map is the canonical projection. From the definition of φ it follows that there is a factorization κ of $(\mathrm{can},\ k(\theta)_H)$ as in the diagram (4.58). By construction $\kappa(z) = 1 \otimes z$ and κ maps into cycles of A.

We show that in homology κ factorizes through proj to a splitting of p as claimed in the statement of the theorem. It suffices to show that elements $z \in J = \ker \mathrm{proj}$ are mapped into boundaries in A. Now J is generated by elements of the form

$$z = c \otimes 1 - 1 \otimes i^*c \quad \text{for} \quad c \in I(G)$$

and

$$\kappa(z) = 1 \otimes z = 1 \otimes c \otimes 1 - 1 \otimes 1 \otimes i^*c .$$

But $I(G)$ is generated by the transgression $c = \tau x$ for $x \in P_{\mathfrak{g}}$, for which by (5.80)

$$d_A(x \otimes 1 \otimes 1) = 1 \otimes c \otimes 1 - 1 \otimes 1 \otimes i^*c$$

and therefore $\kappa\ (z) = d_A(x)$ is a boundary on a set of generators of J.

Via composition with φ the map κ has the desired domain and target. That it is a splitting of the exact sequence (4.53) follows from $\theta \circ i = id : \underline{\underline{h}} \longrightarrow \underline{\underline{h}}$ and the consequence $W(i) \circ k(\theta) = id : W(\underline{\underline{h}}) \longrightarrow W(\underline{\underline{h}})$.

By construction it follows then immediately that $\Delta_* \circ \kappa$ is the Chern-Weil homomorphism of the H-bundle P' extended to the G-bundle P. This finishes the proof of theorem 4.52. □

We illustrate the preceding result by considering an ordinary G-bundle $P \longrightarrow M$. Then $q = n = \dim M$ for the point foliation of M. Any connection ω in P is basic, and therefore the generalized characteristic homomorphism for any H-reduction P' is a map

$$\Delta_* : H^{\cdot}(W(\underline{\underline{g}},H)_{[n/2]}) \longrightarrow H_{DR}(M).$$

We can in particular choose a connection in P' and extend it to a connection in P. By theorem 4.52 it follows that $\Delta_*|H(K_{[n/2]}) = 0$ and in fact $\Delta_* \circ \kappa$ is induced from

$$h_*(P') : I(H) \longrightarrow H_{DR}(M).$$

In spite of the fact that $H(K_{[n/2]})$ need not be trivial, nothing of interest can be gained from Δ_*, since the two geometric structures considered are compatible.

4.59 FUNCTORIALITY OF Δ_*. We discuss the functoriality in (G,H) of the generalized characteristic homomorphism Δ_*, in more detail than during the proof of theorem 4.43.

We begin with a homomorphism $\rho : G \longrightarrow G'$ and the extension

$$\rho_* P \equiv P' = P \times_G G'$$

of a G-bundle P to a G'-bundle P' via ρ. We need to describe explicitly how a connection ω in P determines a connection $\rho_*\omega \equiv \omega'$ in P'. To do this we define a $\underline{\underline{g}}'$-valued form on $P \times G'$, which is G-basic under the diagonal action of G on $P \times G'$, and hence defines a $\underline{\underline{g}}'$-valued form on $P \times_G G'$ which is the desired connection form $\rho_*\omega = \omega'$.

Let more precisely ω be a $\underline{\underline{g}}$-valued connection form on P characterized by

$$R_g^*\omega = Ad(g^{-1})\omega \quad \text{for} \quad g \in G$$

$$\omega(X^*) = x \qquad \text{for} \quad x \in \underline{\underline{g}},$$

where X^* denotes as usual the fundamental vectorfield on P defined by x. Define a $\underline{\underline{g}}'$-valued 1-form ω' on $P \times G'$ by

(4.60) $$\omega'_{(p,g')}(X,Y) = (Ad(g'^{-1}) \circ d\rho \circ \omega_p)(X) + \theta_{g'}(Y)$$

for X a tangent vectorfield on P and Y a tangent vectorfield on G', where θ denotes the $\underline{\underline{g}}'$-valued Maurer-Cartan form on G'. For the particular vectorfields X^* defined by $x \in \underline{\underline{g}}$ and $Y = Y_L$ the left-invariant vectorfield defined by $y \in \underline{\underline{g}}'$ this formula reads

(4.61) $$\omega'_{(p,g')}(X^*,Y_L) = Ad(g'^{-1})d\rho(x)) + y.$$

It is easily verified that ω' is a connection on the trivial G'-bundle $P \times G' \longrightarrow P$. For this one has to use formula

(4.62) $$R'^*_{g'}\theta = Ad(g'^{-1})\,\theta$$

for the Maurer-Cartan form θ of G' under the right translation by g' on G'. We wish to show that ω' is G-basic for the diagonal G-action on $P \times G'$ defined by

$$(4.63) \qquad (p,g') \cdot g = (pg, \rho(g^{-1})g').$$

We use the notations R for the right action of G on P, R' for the right action of G' on G', $P \times G'$ and the quotient P', and further L' for the left action of G' on G' and $P \times G'$. Then (4.63) reads e.g.

$$(4.64) \qquad (p,g') \cdot g = (R_g, L'_{\rho(g^{-1})})(p,g').$$

Beside the usual notation X^* on P for the fundamental vectorfield defined by $x \in \underline{\underline{g}}$ let Y^* denote the fundamental vectorfield on $P \times G'$ defined by the flow of $R'_{\exp ty}$ for $y \in \underline{\underline{g}}'$. We have then

$$(4.65) \qquad Y^* = (0, Y_L) \quad \text{on} \quad P \times G'.$$

We further need the right-invariant vectorfield Y_R on G' defined by the flow $L'_{\exp ty}$ for $y \in \underline{\underline{g}}'$. The vectorfield defined by the diagonal flow of $\exp tx$ on $P \times G'$ is then by (4.64) clearly

$$(4.66) \qquad (X^*, -Y_R), \quad \text{for} \quad y = d\rho(x).$$

It is now clear that ω' on $P \times G'$ is G-basic under the diagonal G-action if and only if

$$(4.67) \qquad \omega'(X^*, -Y_R) = 0 \quad \text{for} \quad x \in \underline{\underline{g}}, y = d\rho(x)$$

$$(4.68) \qquad (R_g, L'_{\rho(g^{-1})})^* \omega' = \omega' \quad \text{for} \quad g \in G.$$

The first formula follows from the definition (4.60) of ω':

$$\omega'_{(p,g')}(X^*, -Y_R) = \mathrm{Ad}(g'^{-1})(d\rho(x)) - \theta_{g'}(Y_R) = 0,$$

since by (4.62)

$$\theta_{g'}(Y_R) = \theta_{g'}((R'_{g'})_* y) = (R'_{g'}{}^*\theta)_e(y) = Ad(g'^{-1})\theta_e(y) = Ad(g'^{-1})y.$$

Formula (4.68) is equivalent to

$$(4.69) \qquad R^*_g\omega' = L'_{\rho(g)}{}^* \cdot \omega'$$

where the actions on $P \times G'$ are induced from the action of G on P and the left action of G' on G'. The verification is as follows.

$$(R^*_g\omega')(X^*,Y_L)_{(p,e)} = \omega'_{(pg,e)}((R_g)_*X,Y_L)$$

$$= d\rho\ \omega_{pg}(Ad(g^{-1})x) + \theta_e(Y_L) = d\rho(Ad(g^{-1})x) + y,$$

and

$$(L'_{\rho(g)}{}^*\omega')(X^*,Y_L)_{(p,e)} = \omega'_{(p,\rho(g))}(X^*,Y_L) = Ad(\rho(g)^{-1})(d\rho(x)) + y.$$

But $Ad(\rho(g^{-1}))d\rho(x) = d\rho(Ad(g^{-1})x)$, so that (4.69) is indeed verified.

This furnishes the construction of the connection $\omega' = \rho_*\omega$ on $P' = \rho_*P$ as the induced 1-form on $P \times_G G'$.

If P is a foliated bundle and ω an adapted connection, then the connection ω' on P' defines a foliation of P' to which it is adapted. The foliation of the base space M is unchanged.

Next we consider a closed subgroup $H \subset G$ and a section $s : M \longrightarrow P/H$ of $P/H \longrightarrow M$ defining an H-reduction of P. Let further $H' \subset G'$ be a closed subgroup and $\rho : G \longrightarrow G'$ a homomorphism with $\rho(H) \subset H'$. Then a canonical H'-reduction of the G'-bundle $P' = \rho_*P$ is given by the section $s' : M \longrightarrow P'/H'$

of $P'/H' \longrightarrow M$ defined below.

Let $\varphi : P \longrightarrow P' = P \times_G G'$ be the canonical map induced by $p \longrightarrow (p,e)$. It is an equivariance with respect to the homomorphism $\rho : G \longrightarrow G'$, and therefore induces a map $\tilde{\varphi} : P/H \longrightarrow P'/H'$. The following diagram is then commutative

$$\begin{array}{ccc} P & \xrightarrow{\varphi} & P' \\ \downarrow & & \downarrow \\ P/H & \xrightarrow{\tilde{\varphi}} & P'/H' \\ & \hat{\pi} \searrow \quad \swarrow \hat{\pi}' & \\ & M & \end{array} \tag{4.70}$$

where $\hat{\pi}$ is induced by $\pi : P \longrightarrow M$ and $\hat{\pi}'$ by $\pi' : P' \longrightarrow M$. With these notations the section s of $\hat{\pi}$ defines a section s' of $\hat{\pi}'$ by

$$s' = \tilde{\varphi} \circ s. \tag{4.71}$$

With these definitions we claim now that there is a commutative diagram

$$\begin{array}{ccc} W(\underline{\underline{g}}',H') & \xrightarrow{\quad W(d\rho) \quad} & W(\underline{\underline{g}},H) \\ \Big\downarrow k(\omega')_{H'} & & \Big\downarrow k(\omega)_H \\ \Omega(P'/H') & & \Omega(P/H) \\ & s'^{*} \searrow \qquad \swarrow s^{*} & \\ & \Omega(M) & \end{array} \tag{4.72}$$

This is the precise formulation of the functoriality of the characteristic homomorphism Δ_* under the homomorphism

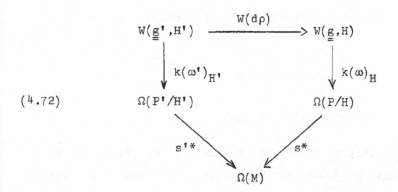

$\rho : (G,H) \longrightarrow (G',H')$. Note that $\Delta_*(P)$ and $\Delta_*(P')$ are induced by the corresponding homomorphisms on the truncated algebras $W(\mathfrak{g},H)_q$ and $W(\mathfrak{g}',H')_q$, and $W(d\rho)$ induces a map $W(\mathfrak{g}',H')_q \longrightarrow W(\mathfrak{g},H)_q$.

Let $\Psi \in W^{s,2r}(\mathfrak{g}',H')$. Then the commutativity of (4.72) translates by 4.46 into the identity

$$(4.73)\quad s^*(W(d\rho)\Psi)(\underbrace{\omega\wedge\ldots\wedge\omega}_{s}\wedge\underbrace{\Omega\wedge\ldots\wedge\Omega}_{r}) = s'^{*}\,\Psi(\underbrace{\omega'\wedge\ldots\wedge\omega'}_{s}\wedge\underbrace{\Omega'\wedge\ldots\wedge\Omega'}_{r})$$

To prove the commutativity of (4.72), it suffices to observe that the diagram

$$(4.74)\qquad \begin{array}{ccc} W(\mathfrak{g}') & \xrightarrow{\;W(d\rho)\;} & W(\mathfrak{g}) \\ \Big\downarrow{\scriptstyle k(\omega')} & & \Big\downarrow{\scriptstyle k(\omega)} \\ \Omega(P') & \xrightarrow{\;\varphi^*\;} & \Omega(P) \end{array}$$

is commutative. The rest follows then from the definition (4.71) of s' and the diagram (4.70). On the generators $\alpha' \in \Lambda^1\mathfrak{g}'$ and $\tilde{\alpha} \in S^1\mathfrak{g}'^*$ the commutativity of (4.74) reduces to the formulas

$$\varphi^*(\alpha'\omega') = (\alpha' \circ d\rho)(\omega)$$

$$\varphi^*(\alpha'\Omega') = (\alpha' \circ d\rho)(\Omega)$$

These are immediately verified from the definition (4.60) of ω' and the fact that $\varphi : P \longrightarrow P'$ is induced by the map $P \longrightarrow P \times G'$ given by $p \longrightarrow (p,e)$. This finishes the proof of the functoriality of the generalized characteristic homomorphism under homomorphisms $(G,H) \longrightarrow (G',H')$.

4.75 RIGIDITY. Let $f: M \longrightarrow X$ be a submersion. An involutive subbundle $L \subset T(f)$ can be considered as a deformation of foliations L_X on the fibers $M_x = f^{-1}(x)$, $x \in X$. Similarly a G-bundle $P \longrightarrow M$ foliated with respect to L defines a deformation of foliated bundles $P_x \longrightarrow M_x$, and an H-structure on P defines a deformation of H-structures on P_x, $x \in M$. The effect on the characteristic homomorphism Δ_* has been discussed in [KT 7], section 8.7. We wish here only to discuss the simplest case of a projection $M = X \times F \longrightarrow X$ with $\dim X = m \geqq 1$. Let q denote the codimension of L_x in the fiber $M_x = F$. Then $q + m$ is the codimension of L in M. The functoriality of Δ_* implies then that for every $x \in X$ the following diagram is commutative

$$\begin{array}{ccc} H(W(\underline{\underline{g}},H)_{q+m}) & \xrightarrow{\Delta_*(P)} & H_{DR}(M) \\ \Big\downarrow \text{can} & & \Big\downarrow \text{restr} \\ H(W(\underline{\underline{g}},H)_q) & \xrightarrow{\Delta_*(P_x)} & H_{DR}(M_x) \end{array} \qquad (4.76)$$

It follows that the classes $\Delta_*(P_x)(u)$ for u in the image of the canonical map $H(W(\underline{\underline{g}},H)_{q+m}) \longrightarrow H(W(\underline{\underline{g}},H)_q)$ induced by projection are rigid for $m \geqq 1$. For a general form of this statement we refer to [KT 7], theorem 8.9. We only wish to point out that this result is an elementary consequence of the functoriality of Δ_*, and does not require any computation of $H(W(\underline{\underline{g}},H)_q)$ for its proof. This generalizes a result of Heitsch [HT] on 1-parameter families of foliations. See also Lehmann [LN 1] for the discussion of deformations.

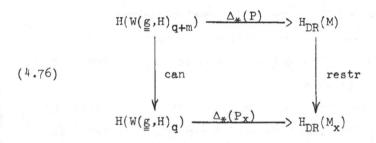

4.77 COMPARISON OF Δ_* WITH CLASSIFYING MAP. In this section we wish to compare the characteristic classes obtained from $\Delta_*(P)$ with those obtained via the classifying maps of P and P'.

Let first $P \longrightarrow M$ be a G-bundle and $g : M \longrightarrow B_G$ its classifying map into the classifying space B_G of G. Then P is the pullback $g^*\eta_G$ of the universal G-bundle $\eta_G = (E_G \longrightarrow B_G)$. The Chern-Weil homomorphism $h_*(P) : I(G) \longrightarrow H_{DR}(M)$ is by functoriality the composition

$$h_*(P) = g^* \circ h_*(\eta_G)$$

where $h_*(\eta_G) : I(G) \longrightarrow H(B_G)$ is the Chern-Weil homomorphism of η_G. (The cohomology discussed here is always the cohomology with real coefficients; the questions of infinite-dimensionality can be avoided by restricting to finite-dimensional skeletons).

Let $\theta : \underline{\underline{g}} \longrightarrow \underline{\underline{h}}$ be an H-equivariant splitting of the exact H-module sequence $0 \longrightarrow \underline{\underline{h}} \longrightarrow \underline{\underline{g}} \longrightarrow \underline{\underline{g}}/\underline{\underline{h}} \longrightarrow 0$. There is then the induced homomorphism κ of (4.53).

On the other hand the H-reduction P' is classified by a map $g' : M \longrightarrow B_H$ and its Chern-Weil homomorphism $h_*(P')$ is the composition

$$h_*(P') = g'^* \circ h_*(\eta_H)$$

where $\eta_H = (E_H \longrightarrow B_H)$ is the universal H-bundle. The maps just discussed fit together with $\Delta_*(P)$ into the following commutative diagram explaining the relation between these maps.

(4.78)

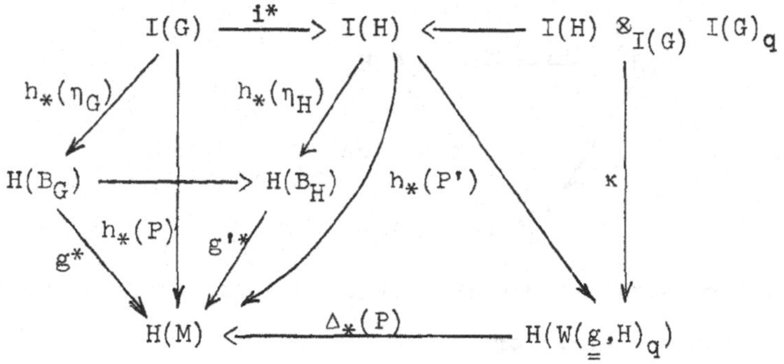

What is needed for an effective use of this construction is the computation of $H(W(\underline{\underline{g}},H)_q)$, which plays the role of an universal algebra of generalized characteristic invariants. Before we turn to this problem in the next chapter, we wish to discuss (4.78) for the special case of a flat bundle and describe a situation where non-trivial elements occur in the image of Δ_*.

For a flat G-bundle P, the Chern-Weil homomorphism $h_*(P)$ is trivial. But g^* need not be trivial (see examples in KT[1] and below). The structure theorem for flat bundles discussed in 2.2 (see [KT 1], proposition 3.1) implies the factorization

(4.79) $$g^* : H(B_G) \xrightarrow{B\alpha^*} H(B_\Gamma) \xrightarrow{\zeta^*} H_{DR}(M)$$

where Γ is the fundamental group of M, $\zeta : M \longrightarrow B_\Gamma$ classifies the universal covering $\tilde{M} \longrightarrow M$, $\alpha : \Gamma \longrightarrow G$ is the holomomy homomorphism and $B\alpha : B_\Gamma \longrightarrow B_G$ the induced map of classifying spaces.

In the flat case $q = 0$ and the map

(4.80) $$\kappa = h(G,H) : I(H) \longrightarrow W(\underline{\underline{g}},H)_0 \cong (\Lambda \underline{\underline{g}}^*)_H$$

has the following interpretation. It is the Chern-Weil homomorphism of the H-bundle G with values in the invariant forms on G/H.

There is a canonical factorization

(4.81)

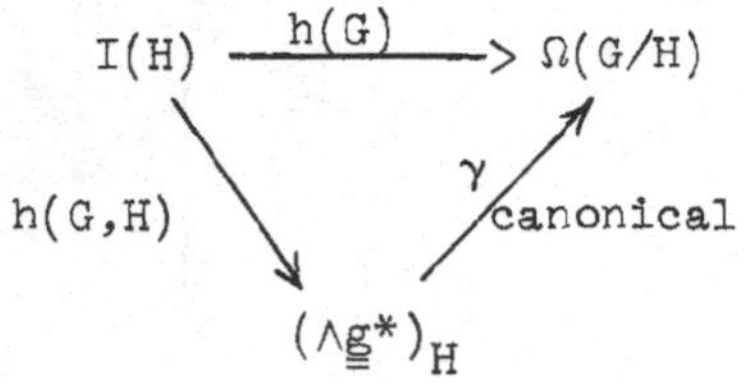

The commutative diagram takes then for a flat G-bundle P the following special form

(4.82)

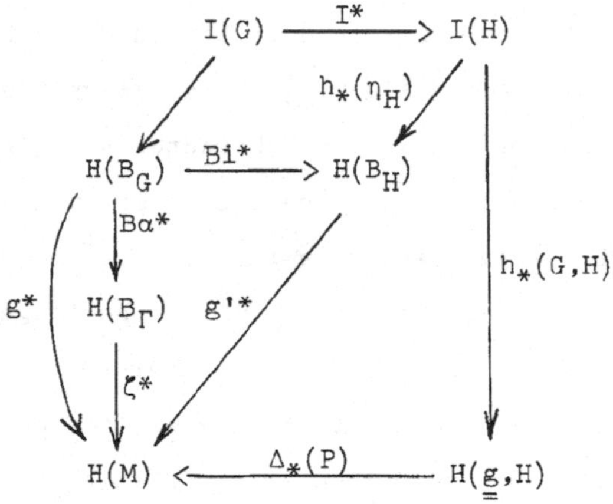

We apply this to the flat bundles considered in [KT 1, 4.14]. Let $K \subset G$ be a maximal compact subgroup of a Lie group G and consider the flat G-bundle

(4.83) $$G \times_K G \cong G/K \times G \longrightarrow G/K.$$

The flat structure is induced by the diagonal action of G (see 2.44). This bundle is obviously the canonical G-extension of the K-bundle $G \longrightarrow G/K$, which hence is a K-reduction of the flat G-bundle. Let $\Gamma \subset G$ be a discrete uniform subgroup operating properly discontinuously and without fixed points on G/K, so that the double coset space $\Gamma \backslash G/K$ is a manifold. By Borel [BO 2] such a Γ exists if G is connected semi-simple with finite center and no compact factor. The flat G-bundle (see 2.50)

$$(4.84) \qquad P = (\Gamma\backslash G) \times_K G \cong G/K \times_\Gamma G \longrightarrow M = \Gamma\backslash G/K$$

on the Clifford-Klein form $\Gamma\backslash G/K$ of the non-compact symmetric space G/K is then canonically equipped with a reduction to the K-bundle $\Gamma\backslash G \longrightarrow \Gamma\backslash G/K$. The characteristic map

$$(4.85) \qquad \Delta_*(P) : H(\underline{\underline{g}},K) \longrightarrow H_{DR}(M)$$

is then by 4.43 well-defined. Since G/K is contractible, it follows that the Clifford-Klein form M is a classifying space B_Γ for the discrete group $\Gamma : M \simeq B_\Gamma$. Therefore $H(M) \cong H(B_\Gamma)$, which is the cohomology $H(\Gamma)$ of the discrete group Γ. Now for compact K, the map $h_*(\eta_K) : I(K) \longrightarrow H(B_K)$ is an isomorphism [BO 1] [C 2]. From (4.82) we obtain the following commutative diagram ([KT 1], (4.18))

$$(4.86) \qquad \begin{array}{ccc} H(B_G) & \xrightarrow{\ \cong\ } & H(B_K) \\ \downarrow{\scriptstyle B\alpha^*} & \swarrow{\scriptstyle g'^*} & \downarrow{\scriptstyle h_*(G,K)} \\ H(\Gamma) \cong H(M) & \xleftarrow{\ \Delta_*(P)\ } & H(\underline{\underline{g}},K) \end{array}$$

The isomorphism $H(B_G) \xrightarrow{\cong} H(B_K)$ is a consequence of the homotopy equivalence $K \simeq G$ which induces a homotopy equivalence $B_K \simeq B_G$. Note that in this case the map $\Delta_*(P)$ is induced by the canonical inclusion

$$(\Lambda^\cdot \underline{\underline{g}}^*)_K \longrightarrow \Omega^\cdot(\Gamma\backslash G/K)$$

so it is really a tautological map. The point of diagram (4.86) is that it relates the existence of non-trivial classes under $\Delta_*(P)$ with the existence of non-trivial classes under the map $B\alpha^*$

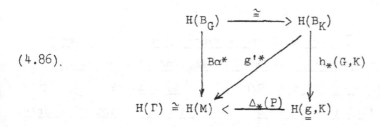

induced by the classifying map $B\alpha : M \longrightarrow B_G$. We have the following result.

4.87 THEOREM [KT 6,7]. *Let G be a connected semi-simple Lie group with finite center and containing no compact factor, $K \subset G$ a maximal compact subgroup and $\Gamma \subset G$ a discrete, uniform and torsion-free subgroup. Then the generalized characteristic homomorphism*

$$\Delta_* : H(\underline{\underline{g}},K) \longrightarrow H_{DR}(M)$$

of the flat bundle

$$P = \Gamma\backslash G \times_K G \cong G/K \times_\Gamma G \longrightarrow \Gamma\backslash G/K$$

is injective.

The proof is based on the following result (see e.g. [KT 1], lemma 4.21).

4.88 LEMMA. *Let $\Gamma \subset G$ be a discrete subgroup of a connected Lie group G operating properly discontinuously and without fixed points on the homogeneous space G/H of G by a closed subgroup $H \subset G$, and such that the manifold $\Gamma\backslash G/H$ is compact and orientable. If $H(\underline{\underline{g}},H)$ satisfies Poincaré duality with respect to a non-zero $\mu \in (\Lambda^n \underline{\underline{g}}^*)_H$, $n = \dim \underline{\underline{g}}/\underline{\underline{h}}$, then the canonical inclusion $\gamma : (\Lambda \underline{\underline{g}}^*)_H \longrightarrow \Omega(\Gamma\backslash G/H)$ induces an injective homomorphism*

$$\gamma_* : H(\underline{\underline{g}},H) \longrightarrow H_{DR}(\Gamma\backslash G/H).$$

Proof. μ defines a G-invariant nowhere zero n-form on G/H, which is a fortiori Γ-invariant and induces hence a volume form on the compact manifold $\Gamma\backslash G/H$. It follows that γ_* is an isomorphism in dimension n.

Let now $x \in H^i(\underline{\underline{g}},H)$ be a non-zero element. By Poincaré-duality there exists $y \in H^{n-i}(\underline{\underline{g}},H)$ such that the cup-product $x.y = \mu \in H^n(\underline{\underline{g}},H)$. Since γ_* is multiplicative and $\gamma_*(x.y) = \gamma_*\mu \neq 0$, it follows that $\gamma_* x \neq 0$. □

Note that for a compact subgroup H the isotropy representation of H in $\underline{\underline{g}}/\underline{\underline{h}}$ is unimodular, so that the existence of a non-zero $\mu \in (\Lambda^n \underline{\underline{g}}^*)_H$ is then always guaranteed. This proves theorem 4.87.

To give a more geometric interpretation of the classes so obtained we proceed as in chapter 4 of [KT 1]. Let $G_{\mathbb{C}}$ be the complexification of G and $U \subset G_{\mathbb{C}}$ a maximal compact subgroup. Then

$$H(\underline{\underline{g}},K) \cong H(\underline{\underline{u}},K) \cong H(U/K)$$

so that the elements of $H(\underline{\underline{g}},K)$ can be realized by cohomology classes of the compact space U/K (whereas G/K is contractible). A typical example is $G = SL(n,\mathbb{R})$ with complexification $SL(n,\mathbb{C})$ and maximal compact subgroup $SU(n) \subset SL(n,\mathbb{C})$. In this case $K = SO(n)$. The map $\Delta_*(P)$ is then realized on the cochain level by the map

$$\gamma\colon (\Lambda \underline{\underline{u}}^*)_K \longrightarrow \Omega(\Gamma\backslash G/K)$$

which is nothing but Matsushima's map constructed in [MT]: an invariant form on U/K is characterized by an element in $(\Lambda \underline{\underline{u}}^*)_K$, which canonically defines an element in $(\Lambda \underline{\underline{g}}^*)_K$, which in turn defines a G-invariant form on G/K, hence a form in $\Omega(\Gamma\backslash G/K)$. Since both forms we start from and end up with are harmonic, the map is really the same as the induced map on the cohomology level, and the injectivity in cohomology is obvious. (In [MT] Matsushima is concerned with the surjectivity problem).

In this special context the map $h_*(G,K)$ in (4.86) can be replaced by the restriction i^* to the fiber of the canonical fibration

$$U/K \xrightarrow{i} B_K \longrightarrow B_U$$

Diagram (4.86) above takes now the form of (4.18) in KT[1]:

$$(4.89) \qquad \begin{array}{ccc} H(B_G) & \xrightarrow{\cong} & H(B_K) \\ \downarrow{\scriptstyle B\alpha^*} & \swarrow{\scriptstyle g'^*} & \downarrow{\scriptstyle i^*} \\ H(\Gamma) \cong H(M) & \xleftarrow{\Delta_*(P)} & H(U/K) \end{array}$$

Note that g' classifies the K-bundle $K \longrightarrow \Gamma\backslash G \longrightarrow \Gamma\backslash G/K$, and i classifies the K-bundle $K \longrightarrow U \longrightarrow U/K$. The factorization

$$(4.90) \qquad g'^* = \Delta_*(P) \circ i^*$$

gives an interpretation of $\Delta_*(P)$ as a proportionality map in the following sense. For any K-module there are associated vectorbundles on $\Gamma\backslash G/K$ and on U/K. The corresponding characteristic numbers of these vectorbundles take values on the respective

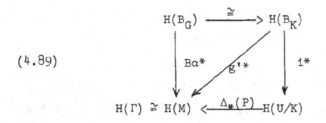

5. COHOMOLOGY OF $\underline{\underline{g}}$-DG-ALGEBRAS

5.0 OUTLINE. In this chapter we give an algorithm for the computation of the cohomology algebras $H(W(\underline{\underline{g}},H)_q)$. These are the algebras which appeared in the construction of the generalized characteristic homomorphisms as algebras of universal generalized characteristic classes.

What we do is to construct a DG-algebra $A(W(\underline{\underline{g}})_q,H)$ and a homomorphism of DG-algebras

$$\varphi = \varphi(W(\underline{\underline{g}})_q,H) : A(W(\underline{\underline{g}})_q,H) \longrightarrow W(\underline{\underline{g}},H)_q$$

which is an isomorphism in homology. If

$$\Delta(\omega) : W(\underline{\underline{g}},H)_q \longrightarrow \Omega(M)$$

denotes the generalized characteristic homomorphism on the cochain level, defined by an adapted connection ω on a foliated G-bundle equipped with an H-reduction, then the composition

$$\tilde{\Delta}(\omega) = \Delta(\omega) \circ \varphi : A(W(\underline{\underline{g}})_q,H) \longrightarrow \Omega(M)$$

is the realization of $\Delta(\omega)$ on the complex $A(W(\underline{\underline{g}})_q,H)$. This is the work announced in [KT 5].

In this chapter we adopt for the sake of simplicity mostly a Lie algebraic point of view, neglecting questions of group actions. To apply these results to geometric situations, we need to make appropriate connectedness assumptions on G and H. The construction of φ is then done in the following more general context. We consider $\underline{\underline{g}}$-DG-algebras E with connections (definition 5.11) satisfying certain finiteness conditions (5.78). The pair $(\underline{\underline{g}},\underline{\underline{h}})$ is assumed to be reductive. We define then a DG-algebra $A(E,\underline{\underline{h}})$ and a functorial DG-homomorphism

$$\varphi(E,\underline{\underline{h}}) : A(E,\underline{\underline{h}}) \longrightarrow E_{\underline{\underline{h}}}$$

which is an isomorphism in homology. For $E = W(\underline{\underline{g}})_q$ this leads to the result mentioned first.

We outline now the topics discussed. The chapter begins with a few generalities, in particular connections and filtrations in $\underline{\underline{g}}$-DG-algebras. Next follows the construction of a semi-simplicial model $W_1(\underline{\underline{g}})$ for the Weil algebra and a map $\lambda : W(\underline{\underline{g}}) \longrightarrow W_1(\underline{\underline{g}})$ (5.34). This construction is a generalization of the Weil lemma establishing the independence of the characteristic homomorphism from the choice of a connection. This construction is fundamental for several reasons. One is the construction of a universal homotopy $\lambda^1 : W(\underline{\underline{g}}) \longrightarrow W(\underline{\underline{g}}) \otimes W(\underline{\underline{g}})$ between the universal connections ε_0 and ε_1 (corollary 5.54). This homotopy preserves the natural actions of $\underline{\underline{g}}$ and gives rise to homotopies between the Weil homomorphisms of any two connections (proposition 5.58). On invariant polynomials this gives in particular a universal form of the Weil lemma (5.58'). Another important application of λ^1 is the construction of a universal transgressive operator (5.70). This construction is of the type of the constructions by Chern-Simons [CS1], which was one of the motivations of our work. This leads finally to the construction of the DG-algebra $A(E,\underline{\underline{h}})$ (5.79) (5.80) and the DG-homomorphism $\varphi(E,\underline{\underline{h}})$ as explained above, and in particular to the computation of $H(W(\underline{\underline{g}},H)_q)$ (theorem 5.110). The result of the evaluation of the characteristic homomorphism on the cochain level $\tilde{\Delta}(\omega) = \Delta(\omega) \circ \varphi$ is theorem 5.95. In the computation of $H(E_{\underline{\underline{h}}})$ we introduce in fact a subcomplex $\hat{A}(E)$ of $A(E,\underline{\underline{h}})$ (5.104). The computation of $H(E_{\underline{\underline{h}}})$ is given in theorem 5.107. Then the characteristic homomorphism $\tilde{\Delta}(\omega)$ has to be evaluated on $\hat{A}(W(\underline{\underline{g}})_q)$. This construction is similar to the construction of a characteristic homomorphism in [B 3] by Bott and Milnor.

Observe however that e.g. the boundary formula (5.97) is a consequence of a universal formula in $W(\underline{\underline{g}})$. More details on this material will appear in [KT 11] [KT 12].

5.1 LIE ALGEBRA COHOMOLOGY WITH COEFFICIENTS (Koszul). We need the following generalities on the cohomology of a Lie algebra $\underline{\underline{g}}$ over a groundfield K of characteristic zero. Let $U(\underline{\underline{g}})$ denote the universal enveloping algebra of $\underline{\underline{g}}$ and denote $x \longrightarrow \underline{x}$ the canonical injection $\underline{\underline{g}} \longrightarrow U(\underline{\underline{g}})$. Then

(5.2) $$V_p(\underline{\underline{g}}) = U(\underline{\underline{g}}) \otimes \wedge_p(\underline{\underline{g}}), \quad p \geq 0$$

defines an augmented complex $V.(g) \longrightarrow K \longrightarrow 0$ of free left $U(\underline{\underline{g}})$-modules. The differential $\delta : V_p \longrightarrow V_{p-1}$ is defined by

(5.3) $$\begin{aligned} \delta(u \otimes x_1 \wedge \ldots \wedge x_p) &= \sum_{j=1}^{p} (-1)^{j+1}\, u\, \underline{x}_j \otimes x_1 \wedge \ldots \wedge \hat{x}_j \wedge \ldots \wedge x_p \\ &+ \sum_{i<j} (-1)^{i+j}\, u \otimes [x_i, x_j] \wedge x_1 \wedge \ldots \wedge \hat{x}_i \wedge \ldots \wedge \hat{x}_j \wedge \ldots \wedge x_p \end{aligned}$$

for $u \in U(\underline{\underline{g}})$; $x_1, \ldots, x_p \in \underline{\underline{g}}$. For any $\underline{\underline{g}}$-module A the complex

(5.4) $$C^{\cdot}(\underline{\underline{g}}, A) = \mathrm{Hom}_{U(\underline{\underline{g}})}(V.(\underline{\underline{g}}), A) \cong \mathrm{Hom}_K(\wedge.(\underline{\underline{g}}), A)$$

is equipped with the Chevalley-Eilenberg differential $d_C : C^p(\underline{\underline{g}}, A) \longrightarrow C^{p+1}(\underline{\underline{g}}, A)$ defined by transposition of δ. For $A = K$ this is the Chevalley-Eilenberg complex $\wedge\, \underline{\underline{g}}^*$ considered before. With these notations the cohomology of $\underline{\underline{g}}$ with coefficients in the $\underline{\underline{g}}$-module A is given by

(5.5) $$H^{\cdot}(\underline{\underline{g}}, A) = \mathrm{Ext}^{\cdot}_{U(\underline{\underline{g}})}(K, A) \cong H(C^{\cdot}(\underline{\underline{g}}, A))$$

If A is a $\underline{\underline{g}}$-algebra, then $C^{\cdot}(\underline{\underline{g}}, A)$ has a multiplication turning $H^{\cdot}(\underline{\underline{g}}, A)$ into an algebra.

It is easy to verify that on the $\underline{\underline{g}}$-invariant elements $C(\underline{\underline{g}},A)^{\underline{\underline{g}}}$ the Chevalley-Eilenberg differential satisfies $d_C = -d_o$, where d_o is obtained from the trivial $\underline{\underline{g}}$-action on A. It follows that for a trivial $\underline{\underline{g}}$-module A we have $d_C = -d_o = 0$ on $C(\underline{\underline{g}},A)^{\underline{\underline{g}}}$.

For a systematic discussion, we need first the right $U(\underline{\underline{g}})$-action on $\wedge\underline{\underline{g}}$ given by

$$(5.6) \qquad (x_1 \wedge \ldots \wedge x_p).y = -\sum_{j=1}^{p} x_1 \wedge \ldots \wedge [y,x_j] \wedge \ldots \wedge x_p$$

for $x_1,\ldots,x_p \in \underline{\underline{g}}$ and $y \in \underline{\underline{g}}$. Then the twisted tensor multiplication in $V.(\underline{\underline{g}}) = U(\underline{\underline{g}}) \otimes \wedge.(\underline{\underline{g}})$ is characterized by the following properties.

(5.7) (i) On $U(\underline{\underline{g}})$, $\wedge\underline{\underline{g}}$ the multiplication is the usual one;

(ii) $(u\otimes 1)\,(1\otimes v) = u\otimes v$ for $u \in U(\underline{\underline{g}})$, $v \in \wedge\underline{\underline{g}}$;

(iii) $(u \otimes v)\,(\underline{y}\otimes 1) = (u\otimes v).y = u\underline{y}\otimes v + u\otimes v.y$ for $u \in U(\underline{\underline{g}})$, $v \in \wedge(\underline{\underline{g}})$ and $y \in \underline{\underline{g}}$. In particular $(1\otimes x)\,(\underline{y}\otimes 1) = \underline{y}\otimes x - 1\otimes[y,x]$ and hence

$$(5.8) \qquad (\underline{y}\otimes 1)\,(1\otimes x) - (1\otimes x)\,(\underline{y}\otimes 1) = 1\otimes[y,x]$$

The differential δ is a derivation of degree -1 with respect to this multiplication.

Let now $E^\bullet$ be a $\underline{\underline{g}}$-DG-algebra (definition 3.13). The operator $\theta(x)$, $i(x)$ for $x \in \underline{\underline{g}}$ give rise to a map $V_p \otimes E^\bullet \longrightarrow E^{\bullet - p}$ characterized by $(p = 1)$:

$$(5.9) \qquad \begin{aligned} (\underline{x}\otimes 1).a &= \theta(x)a \\ (1\otimes x).a &= i(x)a \end{aligned}$$

The $\underline{g}$-DG-algebra structure of E is then equivalently described as a $V(\underline{g})$-module structure on E, such that $\theta(x)$ and $i(x)$ as defined in (5.9) act as derivations on E, and such that the property

$$d(v.a) = \delta v.a + (-1)^{\deg v}\, v.da \tag{5.10}$$

holds for $v \in V_p$, $a \in E^n$ and $v.a \in E^{n-p}$. This last rule is expressed by saying that the $V(\underline{g})$-algebra E is a differential $V(\underline{g})$-algebra. More generally a differential $V(\underline{g})$-module is a $V(\underline{g})$-module with a differential d satisfying (5.10).

In this terminology of Koszul a $\underline{g}$-DG-homomorphism $E^{\cdot} \longrightarrow E'^{\cdot}$ of $\underline{g}$-DG-algebras is the same as a homomorphism of differential $V(\underline{g})$-algebras. E.g. a connection $\omega : \wedge^{\cdot}\underline{g}^* \longrightarrow \Omega^{\cdot}(P)$ in a G-bundle P is a $V(\underline{g})$-algebra homomorphism, and its Weil homomorphism $k(\omega) : W^{\cdot}(g) \longrightarrow \Omega^{\cdot}(P)$ a homomorphism of differential $V(\underline{g})$-algebras.

5.11 CONNECTIONS IN $\underline{g}$-DG-ALGEBRAS. Following Cartan [CA] we adopt the

DEFINITION. _A connection in a commutative_ $\underline{g}$_-DG-algebra_ $E^{\cdot}$ _is a multiplicative homomorphism_ $\omega : \wedge^{\cdot}\underline{g}^* \longrightarrow E^{\cdot}$ of degree 0 which is a $V(\underline{g})$-algebra homomorphism.

By the argument of lemma 4.13 such a map defines a unique homomorphism of differential $V(\underline{g})$-algebras $k(\omega) : W^{\cdot}(\underline{g}) \longrightarrow E^{\cdot}$ (the Weil homomorphism of ω) making the diagram

(5.12)

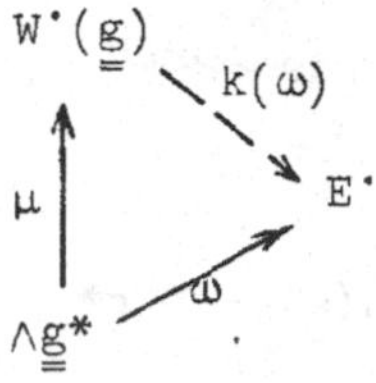

commutative. The canonical map $\mu : \wedge g^* \longrightarrow W(g)$ is a universal connection (extending to the identity on $W(g)$). We shall use the term connection indiscriminately for the connection map ω and its Weil homomorphism $k(\omega)$. Thus e.g. $id : W(g) \longrightarrow W(g)$ is the universal connection in $W(g)$.

5.13 FILTRATIONS IN g-DG-ALGEBRAS. The canonical (Koszul) filtration of a g-DG-algebra $E^\bullet$ is defined by

(5.14) $$F^pE^n = \{a \in E^n \mid v.a = 0 \text{ for } v \in V_q \text{ with } q > n-p\}$$

This property is equivalently expressed by

$$v.a = 0 \quad \text{for} \quad v \in \wedge_q g \quad \text{with} \quad q > n-p$$

and still equivalently by

$$v.a = 0 \text{ for } v = x_1 \wedge \ldots \wedge x_{n-p+1}, \quad x_i \in g.$$

The following properties of the canonical filtration are important:

(5.15) $$F^pE \supset F^{p+1}E,\ F^oE = E \quad \text{and} \quad F^pE^n = 0 \quad \text{for} \quad p > n;$$

(5.16) $F^pE \subset E$ is a g-DG-ideal;

(5.17) $$F^pE \,.\, F^qE \subset F^{p+q}E;$$

(5.18) $F^pE^p = (E^{i(g)})^p$, i.e. the elements a of degree p such that $i(x)a = 0$ for all $x \in g$.

A homomorphism (of degree zero) of g-DG-algebras $E^\bullet \longrightarrow G^\bullet$ is clearly filtration preserving. More generally a homomorphism $E^\bullet \longrightarrow G^{\bullet-\ell}$ of degree $-\ell$ sends F^pE^n into $F^{p-\ell}G^{n-\ell}$.

For the Weil algebra

(5.19) $$F^{2p-1}\, W(g) = F^{2p}\, W(g) = S^pg^* \,.\, W(g)$$

i.e. the canonical filtration coincides with the filtration previously considered on $W(\mathfrak{g})$.

Associated to the filtration $F^pE^\cdot$ there is a canonical map

$$(5.20) \qquad \alpha_p : F^pE^\cdot \longrightarrow \mathrm{Hom}_{U(\mathfrak{g})}(V_{\cdot - p}(\mathfrak{g}), F^pE^p) \cong C^{\cdot - p}(\mathfrak{g}, (E^{i(\mathfrak{g})})^p)$$

defined by

$$\alpha_p(a)v = (-1)^{n(n-p)}\, v.a \quad \text{for} \quad a \in F^pE^n,\ v \in V_{n-p}(\mathfrak{g}) .$$

Clearly $\ker \alpha_p = F^{p+1}E^\cdot$ and we obtain a commutative diagram of $V(\mathfrak{g})$-module maps

$$(5.21) \qquad \begin{array}{ccccccccc} 0 & \longrightarrow & F^{p+1}E^\cdot & \longrightarrow & F^pE^\cdot & \longrightarrow & G^pE^\cdot & \longrightarrow & 0 \\ & & & & & \searrow^{\alpha_p} & \downarrow{\scriptstyle \overline{\alpha}_p} & & \\ & & & & & & C^{\cdot - p}(\mathfrak{g}, (E^{i(\mathfrak{g})})^p) & & \end{array}$$

where $\overline{\alpha}_p$ is the map induced by α_p on the bigraded algebra $G^\cdot E^\cdot$ given by $G^pE^\cdot = F^pE^\cdot / F^{p+1}E^\cdot$. It is easily seen that $\alpha_p d_E = \pm\, d_C \circ \alpha_p$, so that $\overline{\alpha} = (\overline{\alpha}_p)$ is a homomorphism of differential $V(\mathfrak{g})$-algebras (up to sign).

The point of this construction is the following result of Koszul.

5.22 PROPOSITION. <u>A connection</u> ω <u>in the</u> $\mathfrak{g}$-DG-<u>algebra</u> $E^\cdot$ <u>determines a</u> $V(\mathfrak{g})$-<u>module splitting</u>

$$\omega_p : C^{n-p}(\mathfrak{g}, (E^{i(\mathfrak{g})})^p) \longrightarrow F^pE^n$$

<u>of</u> α_p <u>in</u> (5.21) <u>by</u>

$$\omega_p(\varphi \otimes a) = (-1)^{n(n-p)}\omega(\varphi).a$$

for $\varphi \in \Lambda^{n-p}\underline{\underline{g}}^*$ *and* $a \in (E^{i(\underline{\underline{g}})})^p$.

Proof. ω_p is clearly a $V(\underline{\underline{g}})$-module map for $p \geq 0$. To show $\alpha_p \circ \omega_p = \text{id}$, let $v \in \Lambda_{n-p}\underline{\underline{g}}$. Then

$$(\alpha_p \circ \omega_p)(\varphi \otimes a)v = (-1)^{n(n-p)}\alpha_p(\omega(\varphi)\cdot a)v$$

$$= v.(\omega(\varphi).a) = (v.\omega(\varphi)).a = \omega(v.\varphi).a = \varphi(v).a = (\varphi \otimes a)(v) \quad \square$$

It follows that for a $\underline{\underline{g}}$-DG-algebra $E^{\cdot}$ with connection there is an exact sequence of $\underline{\underline{g}}$-DG-algebras (i.e. differential $V(\underline{\underline{g}})$-algebras)

$$(5.23) \quad 0 \longrightarrow F^{p+1}E^{\cdot} \longrightarrow F^pE^{\cdot} \underset{\omega_p}{\overset{\alpha_p}{\rightleftarrows}} C^{\cdot -p}(\underline{\underline{g}},(E^{i(\underline{\underline{g}})})^p) \longrightarrow 0$$

which is split as an exact sequence of $V(\underline{\underline{g}})$-modules. In particular it follows for the graded algebra $G^{\cdot}E^{\cdot}$

$$(5.24) \qquad G^pE^{\cdot} \cong C^{\cdot -p}(\underline{\underline{g}},(E^{i(\underline{\underline{g}})})^p)$$

5.25 COROLLARY. *Let* E *be a* $\underline{\underline{g}}$-DG-*algebra with connection* ω. *Then*

$$F^pE^n = \bigoplus_{r \geq p} \omega(\Lambda^{n-r}\underline{\underline{g}}^*).(E^{i(\underline{\underline{g}})})^r$$

5.26 THE AMITSUR COMPLEX $W_1(\underline{\underline{g}})$ [KT 6]. We consider the Weil algebra

$$W_1^{\ell}(\underline{\underline{g}}) = W(\underline{\underline{g}}^{\ell+1}) = W(\underline{\underline{g}})^{\otimes \ell+1}, \quad \ell \geq 0$$

of the $(\ell+1)$-fold products $\underline{\underline{g}}^{\ell+1} = \underline{\underline{g}} \times \ldots \times \underline{\underline{g}}$. The direct sum

$$(5.27) \qquad W_1(\underline{\underline{g}}) = \bigoplus_{\ell \geq 0} W_1^{\ell}(\underline{\underline{g}}) = \bigoplus_{\ell \geq 0} W(\underline{\underline{g}}^{\ell+1})$$

is itself a $\underline{\underline{g}}$-DG-algebra. At this point we wish only to consider its $\underline{\underline{g}}$-DG-module structure, ignoring the multiplication to which we return in chapter 8.

The operators $i(x)$ and $\theta(x)$ for $x \in \underline{\underline{g}}$ are defined on elements of $W_1^\ell(\underline{\underline{g}})$ via the canonical $\underline{\underline{g}}$-DG-module structure on $W(\underline{\underline{g}})$ ($i(x)$ is affected with a $(-1)^\ell$-sign on $W_1^\ell(\underline{\underline{g}})$).
To explain the differential d_{W_1}, we observe that W_1 is bigraded by

(5.28) $$W_1^{\ell,m} = (W_1^\ell)^m = (W^{\otimes \ell+1})^m$$

The first degree plays the role of a Čech-degree (in a sense to be made clear in Chapter 8). Then a differential of total degree 1 is given by

(5.29) $$d_{W_1} = \delta + (-1)^\ell d_{W_1^\ell}$$

on elements of W_1^ℓ. To explain this, let $\varepsilon_j^\ell : \underline{\underline{g}}^{\ell+2} \longrightarrow \underline{\underline{g}}^{\ell+1}$ be the map defined by

$$\varepsilon_j^\ell(x_0,\ldots,x_{\ell+1}) = (x_0,\ldots,x_{j-1},x_{j+1},\ldots,x_{\ell+1})$$

and $\varepsilon_j^\ell = W(\varepsilon_j^\ell) : W(\underline{\underline{g}}^{\ell+1}) \longrightarrow W(\underline{\underline{g}}^{\ell+2})$ the induced map. Then

(5.30) $$\delta = \sum_{j=0}^{\ell+1} (-1)^j \varepsilon_j^\ell : W_1^\ell \longrightarrow W_1^{\ell+1}.$$

The differential $d_{W_1^\ell} : W_1^{\ell,m} \longrightarrow W_1^{\ell,m+1}$ is the differential in the Weil algebra $W_1^\ell = W^{\otimes \ell+1}$. Thus δ is of bidegree (1,0) and $d_{W_1^\ell}$ of bidegree (0,1).

The canonical projection

(5.31) $$\rho_1 : W_1(\underline{\underline{g}}) \longrightarrow W(\underline{\underline{g}}) \equiv W_1^0(\underline{\underline{g}})$$

mapping $W_1^\ell(\underline{\underline{g}})$ to zero for $\ell > 0$ is then compatible with the $\underline{\underline{g}}$-DG-structure.

We define finally a filtration on $W_1(\underline{\underline{g}})$. For any $m > 0$ the $\underline{\underline{g}}$-DG-module $W(\underline{\underline{g}}^m)$ has the $\underline{\underline{g}}$-filtration (not the canonical one for $m > 1$)

(5.32) $$F_0^{2p}(\underline{\underline{g}})W(\underline{\underline{g}}^m) = \mathrm{Id}\{W^+(\underline{\underline{g}}^m)^{i(\underline{\underline{g}})}\}^{(p)} \text{ (p-th power ideal)}$$

and where $F_0^{2p-1} = F_0^{2p}$. Then we define on $W_1(\underline{\underline{g}})$ a bihomogeneous filtration by $\underline{\underline{g}}$-DG-modules

(5.33) $$F_1^{2p}(\underline{\underline{g}})\, W_1(\underline{\underline{g}}) = \bigoplus_{\ell \geq 0} F_1^{2p}(\underline{\underline{g}})W_1^{\ell}(\underline{\underline{g}}) = \bigoplus_{\ell \geq 0} F_0^{2p}(\underline{\underline{g}})W(\underline{\underline{g}}^{\ell+1})$$

Since $F_0^{2p-1} = F_0^{2p}$, also $F_1^{2p-1} = F_1^{2p}$.

The canonical projection $\rho_1 : W_1 \longrightarrow W$ preserves these filtrations.

5.34 THE MAP $\lambda : W \longrightarrow W_1$. With the preceding notations we can state the following fundamental result [KT 8,12].

5.35 THEOREM. <u>There exists a canonical linear map</u> $\lambda : W^{\cdot}(\underline{\underline{g}}) \longrightarrow W_1^{\cdot}(\underline{\underline{g}})$ <u>of degree zero defined by</u> $\lambda = (\lambda^{\ell})_{\ell \geq 0}$ for maps $\lambda^{\ell} : W^{\cdot}(\underline{\underline{g}}) \longrightarrow W(\underline{\underline{g}}^{\ell+1})^{\cdot-\ell}$ <u>satisfying the following properties:</u>

(5.36)
- (i) λ <u>is a</u> $\underline{\underline{g}}$-DG-<u>module homomorphism;</u>
- (ii) $\rho_1 \circ \lambda = \mathrm{id}$;
- (iii) $\lambda^{\ell}(w) = 0$ <u>for</u> $w \in W^{q,2p}(\underline{\underline{g}})$ <u>and</u> $\ell > p$;
- (iv) $\lambda\, F_0^{2p}(\underline{\underline{g}})W(\underline{\underline{g}}) \subset F_1^{2p}(\underline{\underline{g}})W_1(\underline{\underline{g}})$.

The commutativity of λ with the differentials in W and W_1 reads explicitly

(5.37) $\lambda^{\ell+1} \circ d_W + (-1)^{\ell} d_{W_1^{\ell+1}} \circ \lambda^{\ell+1} = \sum_{j=0}^{\ell+1} (-1)^j \varepsilon_j^{\ell} \circ \lambda^{\ell}, \quad \ell \geq 0.$

We will see in 5.53 how this construction leads for $\ell = 1$ to a homotopy between universal connections. In 5.59 we will discuss applications of this construction.

Proof of theorem 5.35. (i) For every commutative $\underline{\underline{g}}$-DG-algebra E we construct first a simplicial $\underline{\underline{g}}$-DG-algebra $\tilde{E} = (E^{(\ell)})_{\ell \geqq 0}$.

Define

(5.38) $E^{(\ell)} = \Big(E^{\cdot}[t_0,\ldots,t_\ell]/(\sum_{j=0}^{\ell} t_j)-1\Big) \otimes \Big(\wedge(dt_0,\ldots,dt_\ell)/(\sum_{j=0}^{\ell} dt_j)\Big)$

where $t_0,\ldots,t_\ell$ are elements of degree 0 and $dt_0,\ldots,dt_\ell$ elements of degree 1. $E^{(\ell)}$ is considered attached to the standard ℓ-simplex $\Delta^{(\ell)} = \{(t_0,\ldots,t_\ell) \mid \sum_{j=0}^{\ell} t_j = 1 \text{ and } t_j \geqq 0\}$.

The differential $d_{\tilde{E}}$ is defined by sending dt_j to 0 and by the formula

(5.39) $d_{\tilde{E}}(e(t_0,\ldots,t_\ell)) = (d_E e)(t_0,\ldots,t_\ell)$

$$+ (-1)^{\deg e} \sum_{j=0}^{\ell} \frac{\partial}{\partial t_j} e(t_0, \ldots, t_\ell) \otimes dt_j.$$

The face map $\varepsilon_j^{\ell} : \Delta^{(\ell)} \longrightarrow \Delta^{(\ell+1)}$ given by

$$\varepsilon_j^{\ell}(t_0,\ldots,t_\ell) = (t_0,\ldots,t_{j-1},0,t_j, \ldots, t_\ell)$$

induce maps $\varepsilon_j^{\ell} : E^{(\ell+1)} \longrightarrow E^{(\ell)}$ by sending

$t_i;dt_i \longrightarrow t_i,dt_i$ for $0 \leqq i < j$,

$t_j,dt_j \longrightarrow 0$,

$t_i,dt_i \longrightarrow t_{i-1},dt_{i-1}$ for $j < i \leqq \ell + 1$.

This turns $\tilde{E}$ into a simplicial DG-algebra (see chapter 8). Extending i, θ to derivations of degree $-1, 0$ of E^ℓ by $i(t_j) = 0$, $i(dt_j) = 0$, $\theta(t_j) = 0$, $\theta(dt_j) = 0$, this turns $\tilde{E}$ into a simplicial $\underline{g}$-DG-algebra.

A boundary $\partial : E^{(\ell+1)} \longrightarrow E^{(\ell)}$ is defined by

$$\partial = \sum_{j=0}^{\ell+1} (-1)^j \varepsilon_j^\ell . \tag{5.40}$$

Let $\pi_*^\ell : E^{(\ell)\cdot} \longrightarrow E^{\cdot-\ell}$ be defined by integration over the standard simplex $\Delta^{(\ell)}$ as follows. If

$$v \in \left(E^m[t_0,\dots,t_\ell]/(\sum_{j=0}^{\ell} t_j - 1)\right) \otimes dt_1 \wedge \dots \wedge dt_\ell$$

then

$$\pi_*^{(\ell)} v = (-1)^{m\ell} \int_{\Delta^{(\ell)}} v, \tag{5.41}$$

and on $\left(E[t_0,\dots,t_\ell]/(\sum_{j=0}^{\ell} t_j - 1)\right) \otimes \left(\wedge^q(dt_0,\dots,dt_\ell)/(\sum_{j=0}^{\ell} dt_j)\right)$

for $q < \ell$, the map $\pi_*^{(\ell)}$ is defined to be zero. Then $\pi_*^{(\ell)}$ is a $V(\underline{\underline{g}})$-module map. With these definitions one verifies the simplicial Stokes formula

$$\pi_*^{(\ell+1)} \circ d_{E(\ell+1)} + (-1)^\ell d_E \circ \pi_*^{(\ell+1)} = \pi_*^{(\ell)} \circ \partial, \quad \ell \geq 0. \tag{5.42}$$

(ii) Let $E^\cdot$ again be a commutative $\underline{\underline{g}}$-DG-algebra. For any set of $\ell+1$ connections $\omega_j : \wedge \underline{\underline{g}}^* \longrightarrow E^\cdot$ with Weil homomorphisms

$$k(\omega_j) : W^\cdot(\underline{\underline{g}}) \longrightarrow E^\cdot; \quad j = 0,\dots,\ell,$$

let $\sigma = (0,\dots,\ell)$ and consider

(5.43) $$k(\omega^\sigma) : W(\underline{\underline{g}}) \longrightarrow E^{(\ell)},$$

the Weil homomorphism of the connection in $E^{(\ell)}$ determined by

(5.44) $$\omega^\sigma : \underline{\underline{g}}^* \longrightarrow (E^{(\ell)})^1.$$
$$\alpha \longmapsto \sum_{j=0}^{\ell} t_j \omega_j(\alpha)$$

The composition of $k(\omega^\sigma)$ with $\pi_*^{(\ell)}$ defined under (i)

(5.45) $$\lambda_E^\ell(\sigma) = \pi_*^{(\ell)} \circ k(\omega^\sigma) : W^\cdot(\underline{\underline{g}}) \longrightarrow E^{\cdot - \ell}$$

is a $V(\underline{\underline{g}})$-module map of degree $-\ell$.

Let $\varepsilon_j^\ell(\sigma) = (0,\dots,\hat{j},\dots,\ell+1)$. The boundary $\partial\sigma$ is given by

$$\partial\sigma = \sum_{j=0}^{\ell+1} (-1)^j \, \varepsilon_j^\ell(\sigma).$$

The maps $\lambda_E^\ell(\sigma)$ defined for all sets of $\ell+1$ connections, $\ell \geq 0$ satisfy then the following properties:

(5.46)
$$\begin{cases} \text{(i)} & \lambda_E^{\ell+1}(\sigma) \circ d_W + (-1)^\ell d_E \circ \lambda_E^{\ell+1}(\sigma) = \lambda_E^\ell(\partial\sigma); \\ \text{(ii)} & \lambda_E^0(j) = k(\omega_j); \\ \text{(iii)} & \lambda_E^\ell(\sigma)w = 0 \quad \text{for} \quad w \in W^{q,2p}, \quad \ell > p; \\ \text{(iv)} & \lambda_E^1(0,1)\tilde{\alpha} = \alpha(\omega_1 - \omega_0) \quad \text{for} \quad \alpha \in \underline{\underline{g}}^*, \quad \sigma = (0,1), \ \omega = (\omega_0, \omega_1) \end{cases}$$

The proof of (i) follows from the Stokes formula (5.42). Property (ii) is immediate from the definition. To verify (iii), we need to evaluate for an ℓ-simplex σ the curvature in $E^{(\ell)}$

$$k(\omega^\sigma)\tilde{\alpha} = d_{E^{(\ell)}}\omega^\sigma\alpha - \omega^\sigma \, d_\wedge \, \alpha$$

for $\alpha \in \underline{\underline{g}}^*$. If we use $\sum_{j=0}^{\ell} t_j = 1$, then

$$\omega = \omega_0 + \sum_{j=1}^{\ell} t_j(\omega_j - \omega_0)$$

and by (5.39) (3.6)

$$(5.47) \quad k(\omega^\sigma)\tilde{\alpha} = d_E\ \omega^\sigma(\alpha) + \frac{1}{2}\ \alpha[\omega^\sigma,\omega^\sigma] - \sum_{j=1}^{\ell} \alpha(\omega_j - \omega_0) \otimes dt_j.$$

It follows by multiplicativity that for $w \in W^{q,2p}$ with $p < \ell$ property (iii) holds. Note that for $\ell = 1$, $\sigma = (0,1)$, $\omega = (\omega_0,\omega_1)$ we get from (5.47) in particular

$$k(\omega^{(0,1)})\tilde{\alpha} = k(\omega_0)\tilde{\alpha} + t(d\alpha(\omega_1 - \omega_0) + \alpha[\omega_0,\omega_1 - \omega_0]) + \frac{1}{2}\ t^2\alpha[\omega_1 - \omega_0,\omega_1 - \omega_0]$$
$$- \alpha(\omega_1 - \omega_0) \otimes dt$$

It follows that

$$(5.48) \qquad \lambda_E^1(0,1)\tilde{\alpha} = \alpha(\omega_1 - \omega_0)$$

which proves (iv).

Note for later use that in this case (i) reads

$$\lambda_E^1(0,1)d_W + d_E\ \lambda_E^1(0,1) = \lambda_E^0\ (\partial(0,1)) = k(\omega_1) - k(\omega_0)$$

where the last equality follows from (ii). But for $\Phi \in I(\underline{\underline{g}})$ we have $d_W\Phi = 0$ (see chapter 4, argument following 4.16), so that

$$(5.49) \qquad d_E\ \lambda_E^1(0,1)\Phi = (k(\omega_1) - k(\omega_0))\Phi \quad \text{for} \quad \Phi \in I(\underline{\underline{g}}).$$

(iii) Let now more generally $\underset{\sim}{E}$ be a local system of commutative $\underline{\underline{g}}$-DG-algebras on a semi-simplicial set S. We refer to chapter 8, section 8.2 for the precise terminology. Assume that for every 0-simplex $i \in S_0$ there is a connection ω_i given in E_i. For every $\sigma = (i_0,\ldots,i_\ell) \in S_\ell$ there are then maps as explained in part (i) (ii) of this proof

$$\pi_*^{(\ell)}(\sigma) \; : \; E_\sigma^{(\ell)\cdot} \longrightarrow E_\sigma^{\cdot-\ell}$$

$$k(\omega^\sigma) \; : \; W(\underline{\underline{g}}) \longrightarrow E_\sigma^{(\ell)}$$

Note that $k(\omega^\sigma)$ is defined via the connections

$$k(\omega_{i_j}) : W(\underline{\underline{g}}) \longrightarrow E_{i_j} \xrightarrow{\text{can}} E_\sigma, \qquad \sigma = (i_0,\ldots,i_\ell).$$

For every $\sigma \in S_{\ell+1}$ we have then with selfexplanatory notations a commutative diagram

(5.50)

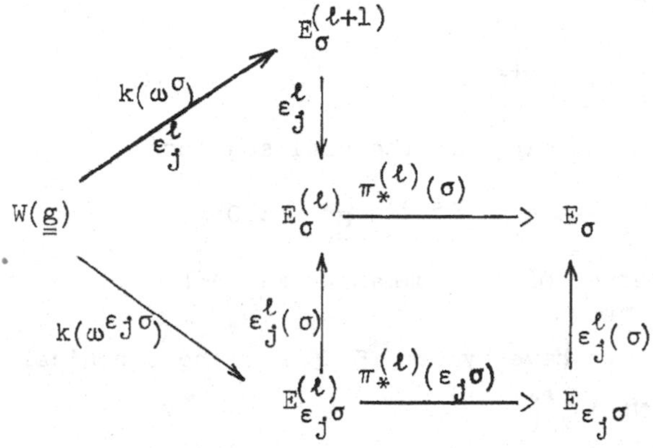

Define then for every $\sigma \in S_\ell$

$$(5.51) \qquad \lambda_{\underset{\sim}{E}}^\ell(\sigma) = \pi_*^{(\ell)}(\sigma) \circ k(\omega^\sigma) \; : \; W^\cdot(\underline{\underline{g}}) \longrightarrow E_\sigma^{\cdot-\ell}$$

By (5.42), which holds for every σ and (5.50), we have then

$$(5.52) \quad \lambda_E^{\ell+1}(\sigma) \circ d_W + (-1)^\ell d_{E_\sigma} \circ \lambda_{\underset{\sim}{E}}^{\ell+1}(\sigma) = \sum_{j=0}^{\ell+1} (-1)^j \varepsilon_j^\ell(\sigma) \lambda_{\underset{\sim}{E}}^\ell(\varepsilon_j^\ell(\sigma)) \equiv \delta\lambda_{\underset{\sim}{E}}^\ell(\sigma)$$

It follows that

$$\lambda^{\cdot}_{\underset{\sim}{E}} = (\lambda^{\ell}_{\underset{\sim}{E}})_{\ell \geq 0} : W^{\cdot}(\underline{\underline{g}}) \longrightarrow C^{\cdot}(S,\underset{\sim}{E}^{\cdot})$$

is a $\underline{\underline{g}}$-DG-module map of degree 0.

As explained in chapter 8, for $S = PT$ (the terminal object) in the category of semi-simplicial sets) and the local system $\underset{\sim}{W}$ given as $W_{\sigma_\ell} = W(\underline{\underline{g}})^{\ell+1}$ for the unique ℓ-simplex σ_ℓ we can write

$$W_1^{\cdot}(\underline{\underline{g}}) = C^{\cdot}(Pt,\underset{\sim}{W}).$$

The preceding construction furnishes then the existence of the map

$$\lambda = (\lambda^{\ell})_{\ell \geq 0} : W(\underline{\underline{g}}) \longrightarrow W_1(\underline{\underline{g}})$$

of theorem 5.35.

Concerning the filtration property (iv) in (5.36) for $\lambda : W \longrightarrow W_1$ we observe the following. Let V_ℓ be defined by the exact sequence

$$0 \longrightarrow \underline{\underline{g}} \xrightarrow{\Delta} \underline{\underline{g}}^{\ell+1} \longrightarrow V_\ell \longrightarrow 0,$$

where Δ is the diagonal map. In the dual sequence

$$0 \longrightarrow V_\ell^* \longrightarrow (\underline{\underline{g}}^*)^{\ell+1} \xrightarrow{\Delta^*} \underline{\underline{g}}^* \longrightarrow 0$$

V_ℓ^* occurs as the kernel of the summation map $\Delta^*(\alpha_0,\ldots,\alpha_\ell) = \sum_{i=0}^{\ell} \alpha_i$.

With these notations we have by Cor. 5.25 for the canonical $\underline{\underline{g}}$-filtration on $W(\underline{\underline{g}}^{\ell+1}) = W(\underline{\underline{g}})^{\otimes \ell+1}$

$$F^p W(\underline{\underline{g}}^{\ell+1})^n \cong \bigoplus_{r=s+2t \geq p} \Lambda^{n-r}\underline{\underline{g}}^* \otimes \{\Lambda^s V_\ell^* \otimes S^t(\underline{\underline{g}}^{\ell+1})^*\},$$

while by (5.32), (5.33) the filtration $F_1(\underline{\underline{g}})$ on $W_1^{\ell}(\underline{\underline{g}}) = W(\underline{\underline{g}}^{\ell+1})$ is given by

$$F_1^{2p}(\underline{\underline{g}})\; W_1^{\ell}(\underline{\underline{g}})^n = \bigoplus_{r=s+t \geq p} \Lambda^{n-r}\underline{\underline{g}}^* \otimes \{\Lambda^s V_\ell^* \otimes S^t(\underline{\underline{g}}^{\ell+1})^*\}.$$

Since $\lambda^{\ell} : W(\underline{\underline{g}})^{\cdot} \longrightarrow W_1^{\ell}(\underline{\underline{g}})^{\cdot-\ell} = W(\underline{\underline{g}}^{\ell+1})^{\cdot-\ell}$ is a $V(\underline{\underline{g}})$-module map of degree $-\ell$ it follows by the remark preceding (5.19) that λ^{ℓ} reduces the canonical filtration-degree by ℓ. Using the above formulas for F^{2p} and $F_1^{2p}(\underline{\underline{g}})$ as well as the definition of λ^{ℓ}, one verifies that λ^{ℓ} satisfies

$$\lambda^{\ell}(F^{2p}(\underline{\underline{g}})W(\underline{\underline{g}})) \subset F_1^{2p}(\underline{\underline{g}})\ W_1^{\ell}(\underline{\underline{g}})$$

for the filtration defined in (5.32) (5.33). This fact is crucial for the characteristic homomorphism as defined in chapter 8 from local data. Proposition 8.22 shows namely that the composition $k_1(\omega) \circ \lambda : W(\underline{\underline{g}}) \longrightarrow \check{C}(\mathcal{U},\pi_*\Omega_P^{\cdot})$ is filtration preserving, which is the basic fact underlying the construction of the characteristic homomorphism Δ_* for foliated bundles. □

5.53 THE UNIVERSAL HOMOTOPY $\lambda^1 : W \longrightarrow W \otimes W$. The properties of the map λ^1 alone in theorem 5.35 are summarized as follows. Let $\varepsilon_i : W(\underline{\underline{g}}) \longrightarrow W(\underline{\underline{g}}) \otimes W(\underline{\underline{g}})$ be the connections defined by $\varepsilon_0(w) = 1 \otimes w$, $\varepsilon_1(w) = w \otimes 1$. Then theorem 5.35 implies the following result.

5.54 COROLLARY. _There exists a canonical_ $V(\underline{\underline{g}})$-_module map of degree_ -1

$$\lambda^1 : W(\underline{\underline{g}}) \longrightarrow W(\underline{\underline{g}}) \otimes W(\underline{\underline{g}})$$

satisfying the following properties:

(i) $\lambda^1 \circ d_W + d_{W\otimes W} \circ \lambda^1 = \varepsilon_0 - \varepsilon_1$, _i.e._ λ^1 _is a homotopy between_ ε_0 _and_ ε_1;

(ii) $\lambda^1 F_0^{2p}(\underline{\underline{g}})\ W(\underline{\underline{g}}) \subset F_1^{2p}(\underline{\underline{g}})\ W(\underline{\underline{g}} \times \underline{\underline{g}})$

for the filtrations F_0 _and_ F_1 _on_ $W(\underline{\underline{g}})$ _and_ $W(\underline{\underline{g}} \times \underline{\underline{g}})$.

The statement that λ^1 is a $V(\underline{\underline{g}})$-module homomorphism reads in this case (see p. 189): for $w \in W(\underline{\underline{g}})$ and $x \in \underline{\underline{g}}$

$$\lambda^1\ i(x)w = -i(x)\ \lambda^1 w$$

$$\lambda^1\ \theta(x)w = \theta(x)\ \lambda^1 w \quad .$$

For later use we evaluate $\lambda^1\tilde{\alpha}$ on an element $\tilde{\alpha} \in S^1\underline{\underline{g}}^*$ corresponding to $\alpha \in \Lambda^1\underline{\underline{g}}^*$. For the connection $\omega_0 = \varepsilon_1$, $\omega_1 = \varepsilon_0$ we have by (5.48) resp. (5.46) (iv) the formula

$$(5.55) \qquad \lambda^1\tilde{\alpha} = \varepsilon_0(\alpha) - \varepsilon_1(\alpha) = 1 \otimes \alpha - \alpha \otimes 1.$$

The interest of this construction is that it provides functorial homotopies between connections as follows. Let $k' : W(\underline{\underline{g}}) \longrightarrow E'$ and $k'' : W(\underline{\underline{g}}) \longrightarrow E''$ be connections in $\underline{\underline{g}}$-DG-algebras E' and E''. Let E be a $\underline{\underline{g}}$-DG-algebra and

$$\mu : E' \otimes E'' \longrightarrow E$$

a pairing which is a $\underline{\underline{g}}$-DG-homomorphism. The $\underline{\underline{g}}$-DG-homomorphisms

$$k_0 = \mu \circ k' \otimes k'' \circ \varepsilon_0 = \mu \circ (1 \otimes k'')$$

$$k_1 = \mu \circ k' \otimes k'' \circ \varepsilon_1 = \mu \circ (k' \otimes 1)$$

are connections in E making the diagrams

$$(5.56) \qquad \begin{array}{ccccc} W(\underline{\underline{g}}) & \xrightarrow{\varepsilon_i} & W(\underline{\underline{g}}) \otimes W(\underline{\underline{g}}) & \xrightarrow{k' \otimes k''} & E' \otimes E'' \\ & \searrow^{k_i} & & & \downarrow \mu \\ & & & & E \end{array}$$

commutative. Then the "difference map" of degree -1

$$\lambda^1(k',k'') = \mu \circ k' \otimes k'' \circ \lambda^1 : W(\underline{\underline{g}}) \longrightarrow E$$

is a homotopy between k_0 and k_1, i.e.

$$(5.57) \qquad \lambda^1(k',k'') \circ d_W + d_E \circ \lambda_1(k',k'') = k_0 - k_1.$$

The proof is a trivial consequence of (5.54), (i). For $E' = E'' = E$

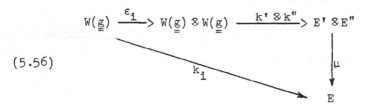

the multiplication in E is a $\mathfrak{g}$-DG-homomorphism. The preceding remarks prove then the following result.

5.58 PROPOSITION. *Let* $E^{\cdot}$ *be a* $\mathfrak{g}$-*DG-algebra with multiplication* μ *and* $k_0, k_1 : W^{\cdot}(\mathfrak{g}) \longrightarrow E^{\cdot}$ *any two connections. Then the* $V(\mathfrak{g})$-*map*

$$\lambda^1(k_0,k_1) = \mu \circ k_0 \otimes k_1 \circ \lambda^1 : W^{\cdot}(\mathfrak{g}) \longrightarrow E^{\cdot -1}$$

is a homotopy between k_0 *and* k_1, *i.e.*

$$\lambda^1(k_0,k_1) \circ d_W + d_E \circ \lambda^1(k_0,k_1) = k_1 - k_0 .$$

The manufacture of homotopies $\lambda^1(k_0,k_1)$ preserving the operators $i(x)$ and $\theta(x)$ is in fact one of the main points of the construction of λ in theorem 5.35.

Since for an invariant polynomial $\Phi \in I(\mathfrak{g})$ the differential $d_W\Phi = 0$, the homotopy formula in proposition 5.58 reduces for $\Phi \in I(\mathfrak{g})^+$ to the formula in $E_{\mathfrak{g}}$:

$$(5.58') \qquad d_E\, \lambda^1(k_0,k_1)(\Phi) = h_1(\Phi) - h_0(\Phi),$$

where as earlier $h = k|I(\mathfrak{g})$ denotes the Chern-Weil homomorphism. This formula is a universal form of the Weil-lemma, stating that the Chern-Weil homomorphism $h_* : I(\mathfrak{g}) \longrightarrow H(E_{\mathfrak{g}})$ is independent of the realizing connection in E.

5.59 APPLICATIONS OF λ^1. Let E be a $\mathfrak{g}$-DG-algebra with connection $k : W(\mathfrak{g}) \longrightarrow E$. Consider the $\mathfrak{g}$-DG-homomorphism $\alpha = (id,k) : E \otimes W(\mathfrak{g}) \longrightarrow E$. Together with $i_0 = 1 \otimes id : W(\mathfrak{g}) \longrightarrow E \otimes W(\mathfrak{g})$ we get the commutative diagram

$$\begin{array}{ccc} E \otimes W(\mathfrak{g}) & \xrightarrow{\ \alpha\ } & E \\ & {\scriptstyle i_0}\nwarrow \quad \nearrow{\scriptstyle k} & \\ & W(\mathfrak{g}) & \end{array}$$

Consider further the $\underline{\underline{g}}$-DG-homomorphism $i_1 = \mathrm{id} \otimes 1 : E \to E \otimes W(\underline{\underline{g}})$. Clearly $\alpha \circ i_1 = \mathrm{id}_E$. The other composition is $i_1 \circ \alpha = (i_1, i_1 \circ k) : E \otimes W(\underline{\underline{g}}) \longrightarrow E \otimes W(\underline{\underline{g}})$. We have then the following result.

5.61 THEOREM. *Let E be a $\underline{\underline{g}}$-DG-algebra with connection $k : W(\underline{\underline{g}}) \longrightarrow E$. There is a homotopy*

$$i_0 \simeq i_1 \circ k : W(\underline{\underline{g}}) \longrightarrow E \otimes W(\underline{\underline{g}})$$

compatible with the operators $i(x)$ and $\theta(x)$ for $x \in \underline{\underline{g}}$. It follows that there is a homotopy

$$\mathrm{id} = (i_1, i_0) \simeq i_1 \circ \alpha : E \otimes W(\underline{\underline{g}}) \longrightarrow E \otimes W(\underline{\underline{g}})$$

compatible with the operators $i(x)$ and $\theta(x)$ for $x \in \underline{\underline{g}}$.

Proof. We apply proposition 5.58 to the $\underline{\underline{g}}$-DG-homomorphisms i_0 and $i_1 \circ k$. Then

$$\lambda^1(i_0, i_1 \circ k) = \mu \circ i_0 \otimes (i_1 \circ k)) \circ \lambda^1 : W^{\cdot}(\underline{\underline{g}}) \to (E \otimes W(\underline{\underline{g}}))^{\cdot -1}$$

is a homotopy as desired. □

5.62 COROLLARY (Cartan [CA]). *Let E be a $\underline{\underline{g}}$-DG-algebra with connection $k : W(\underline{\underline{g}}) \longrightarrow E$. Then the $\underline{\underline{g}}$-DG-homomorphism*

$$\alpha = (\mathrm{id}, k) : E \otimes W(\underline{\underline{g}}) \longrightarrow E$$

induces for any subalgebra $\underline{\underline{h}} \subset \underline{\underline{g}}$ an isomorphism

$$\alpha_* : H((E \otimes W(\underline{\underline{g}}))_{\underline{\underline{h}}}) \xrightarrow{\cong} H(E_{\underline{\underline{h}}}).$$

Proof. This follows from the preceding results, since α has been shown to be a homotopy equivalence compatible with the $\underline{g}$-DG-structures. □

Note that α_* is independent of the connection k in E, since it is the inverse of the cohomology map induced by $i_1 = id \otimes 1 : E \longrightarrow E \otimes W(\underline{g})$.

Another application of the universal homotopy λ^1 is the following result.

5.64 THEOREM. Let $(\underline{g},\underline{h})$ be a pair of Lie algebras and $\theta : \underline{g} \longrightarrow \underline{h}$ an equivariant splitting of the exact $\underline{h}$-module sequence $0 \longrightarrow \underline{h} \xrightarrow{i} \underline{g} \longrightarrow \underline{g}/\underline{h} \longrightarrow 0$. Let $k(\theta) : W(\underline{h}) \longrightarrow W(\underline{g})$ be the Weil homomorphism of the connection $\mu \circ \wedge\theta^* : \wedge\underline{h}^* \longrightarrow \wedge\underline{g}^* \longrightarrow W(\underline{g})$, where $\mu : \wedge\underline{g}^* \longrightarrow W(\underline{g})$ is the canonical map. Then the homomorphisms

$$W(\underline{h}) \underset{W(i)}{\overset{k(\theta)}{\rightleftarrows}} W(\underline{g})$$

are inverse homotopy equivalences compatible with the $\underline{h}$-DG-algebra structures. It follows in particular that the induced map on $\underline{h}$-basic elements

(5.65) $$k(\theta)_{\underline{h}} : I(\underline{h}) \longrightarrow W(\underline{g},\underline{h})$$

is a homotopy equivalence with inverse i^*.

Proof. Since $\theta \circ i = id : \underline{h} \longrightarrow \underline{h}$, it follows immediately that $W(i) \circ k(\theta) = id : W(\underline{h}) \longrightarrow W(\underline{h})$. It remains to show that there is a homotopy

(5.66) $$k(\theta) \circ W(i) \simeq id : W(\underline{g}) \longrightarrow W(\underline{g})$$

compatible with the $\underline{h}$-DG-algebra structure on $W(\underline{g})$ defined by restriction of the operators from $\underline{\underline{g}}$ to $\underline{\underline{h}}$. Consider the $\underline{\underline{h}}$-DG-homomorphisms $k_0 = id_{W(\underline{\underline{g}})}$ and $k_1 = k(\theta) \circ W(i) : W(\underline{\underline{g}}) \longrightarrow W(\underline{\underline{g}})$. The difference map

$$\lambda^1(\theta) \equiv \mu \circ (id \otimes k(\theta) \circ W(i)) \circ \lambda^1 : W(\underline{\underline{g}}) \longrightarrow W(\underline{\underline{g}}) \tag{5.67}$$

of degree-1 is a $V(\underline{\underline{h}})$-homomorphism satisfying by (5.58)

$$\lambda^1(\theta) \circ d_W + d_W \circ \lambda^1(\theta) = \mu \circ (id \otimes k(\theta) \circ W(i)) \ (\varepsilon_0 - \varepsilon_1)$$

$$= k_1 - k_0 = k(\theta) \circ W(i) - id.$$

This proves the desired result. □

5.68 TRANSGRESSIVE OPERATOR, SUSPENSION, PRIMITIVE ELEMENTS. Let

$$\lambda^1 : I(\underline{\underline{g}}) \longrightarrow W(\underline{\underline{g}}) \otimes W(\underline{\underline{g}}))_{\underline{\underline{g}}}$$

be the universal homotopy operator of corollary 5.54 restricted to $\underline{\underline{g}}$-basic elements. Since $d\Phi = 0$ for $\Phi \in I(\underline{\underline{g}})$, by (5.54) (i) it follows that

$$d\lambda^1\Phi = 1 \otimes \Phi - \Phi \otimes 1. \tag{5.69}$$

Let $\kappa : W(\underline{\underline{g}}) \longrightarrow K$ be the augmentation to the groundfield K. Then $id \otimes \kappa : W(\underline{\underline{g}}) \otimes W(\underline{\underline{g}}) \longrightarrow W(\underline{\underline{g}})$ is a map compatible with the operators $\theta(x)$ for $x \in \underline{\underline{g}}$. For the composition

$$T_o = -(id \otimes \kappa) \circ \lambda^1 : W(\underline{\underline{g}}) \longrightarrow W(\underline{\underline{g}}) \tag{5.70}$$

we have then

$$T_o \circ d + d \circ T_o = id - \kappa \tag{5.71}$$

on $W(\underline{\underline{g}})$ and $W(\underline{\underline{g}})^{\underline{\underline{g}}}$, and by (5.69) we have for $\Phi \in I(G)^+$:

$$d \circ T_o(\Phi) = -(\mathrm{id} \otimes \kappa) \circ d\lambda^1\Phi = \Phi. \tag{5.72}$$

A linear map $T : I(G)^+ \longrightarrow W(\underline{\underline{g}})$ of degree -1 such that $dT(\Phi) = \Phi$ for $\Phi \in I(G)^+$ is called a universal transgression operator. Let

$$\pi : W(\underline{\underline{g}}) \longrightarrow \Lambda\underline{\underline{g}}^* = W(\underline{\underline{g}})/F^2W(\underline{\underline{g}})$$

denote the canonical projection and

$$\sigma_T = \pi \circ T : (I(G)^+ \longrightarrow \Lambda\underline{\underline{g}}^* \tag{5.73}$$

the composition with T. We wish to show that σ_T induces a linear mapping

$$\sigma : I^{2p}(G) \longrightarrow H^{2p-1}(\underline{\underline{g}}), \quad p > 0,$$

which is independent of the choice of T. First we have $d_\Lambda\sigma(\Phi) = \pi(dT(\Phi)) = \pi(\Phi) = 0$ for $\Phi \in I(G)^+$. Hence σ_T induces a mapping into $H(\underline{\underline{g}}) = H(\Lambda\underline{\underline{g}}^*)$. Consider two universal transgression operators T, T'. For $\Phi \in I(G)^+$ the difference $(T - T')\Phi$ is closed. Since by (5.71) $W(\underline{\underline{g}})^+$ is a cohomologically trivial algebra, there is an element $\Psi \in W(\underline{\underline{g}})$ such that $(T - T^1)\Phi = d\Psi$. Therefore

$$\sigma_T(\Phi) - \sigma_{T'}(\Phi) = \pi \circ (T - T')\Phi = \pi d(\Psi) = d_\Lambda(\pi\Psi)$$

and σ is well-defined. Observing that $d_\Lambda = 0$ on $(\Lambda\underline{\underline{g}}^*)^{\underline{\underline{g}}}$ it follows by the same argument that for a $\underline{\underline{g}}$-invariant $T : I(G)^+ \longrightarrow W(\underline{\underline{g}})^{\underline{\underline{g}}}$ the mapping $\sigma = \sigma_T : I(G)^+ \longrightarrow (\Lambda\underline{\underline{g}}^*)^{\underline{\underline{g}}}$ is independent of T and realizes σ on the level of invariant cochains. This applies in particular to T_o. In either case the map σ is called the suspension map [CA].

The additive map

$$\sigma : I^{2p}(\underline{\underline{g}}) \longrightarrow (\wedge^{2p-1}\underline{\underline{g}}^*)^{\underline{\underline{g}}}, \; p > 0$$

is evaluated for $\Phi \in I^{2p}(\underline{\underline{g}})$ by the formula [KT 11]

$$(5.74) \qquad \sigma\Phi = (-\tfrac{1}{2})^{p-1}\frac{(p-1)!p!}{(2p-1)!} \; \Phi(\underbrace{[\theta,\theta] \wedge \ldots \wedge [\theta,\theta]}_{p-1 \text{ factors}} \wedge \theta).$$

θ is the identity 1-form on $\underline{\underline{g}}$ (in case of a group G the restriction to $\underline{\underline{g}}$ of the Maurer-Cartan form of G). $[\theta,\theta]$ is defined by $[\theta,\theta](x,y) = 2[\theta(x),\theta(y)] = 2[x,y]$ for $x,y \in \underline{\underline{g}}$. The product $[\theta,\theta] \wedge \ldots \wedge [\theta,\theta] \wedge \theta$ finally is the $S^p\underline{\underline{g}}$-valued form which is the exterior product of the (p-1)-fold product of $[\theta,\theta]$ with itself and a copy of θ with respect to the multiplication in $S^{\cdot}\underline{\underline{g}}^*$.

Next we give an explicit representation of the (2p-1)-form $\sigma\Phi$ on $\underline{\underline{g}}$. Let $x_1, \ldots, x_m$ be a basis of $\underline{\underline{g}}$ with dual basis $x_1^*, \ldots, x_m^*$ of $\underline{\underline{g}}^* = \wedge^1\underline{\underline{g}}^*$. When viewed as elements of $\underline{\underline{g}}^* = S^1\underline{\underline{g}}^*$, these elements are denoted $\tilde{x}_1^*, \ldots, \tilde{x}_m^*$. For $\Phi \in I^{2p}(\underline{\underline{g}})$ we have a representation

$$(5.75) \qquad \Phi = \sum_{j_1 \ldots j_p} a_{j_1 \ldots j_p} \tilde{x}_{j_1}^* \ldots \tilde{x}_{j_p}^*$$

with symmetric coefficients $a_{j_1 \ldots j_p}$. Then the suspension $\sigma\Phi$ is given by the formula

$$(5.76) \qquad \sigma\Phi = \frac{(p-1)!p!}{(2p-1)!} \sum_{j_1 \ldots j_p} a_{j_1 \ldots j_p} \, dx^*_{j_1} \wedge \ldots \wedge dx^*_{j_{p-1}} \wedge x^*_{j_p}$$

where $dx_j^* \in \wedge^2\underline{\underline{g}}^*$ denotes the Chevalley-Eilenberg differential of $x_j^* \in \wedge^1\underline{\underline{g}}^*$. To see this, observe that the formula $d\alpha(x,y) = -\alpha[x,y]$

for $\alpha \in \underline{\underline{g}}^*$ can be written in terms of θ as $d\alpha(x,y) = -\alpha[\theta(x),\theta(y)] = -\frac{1}{2}\alpha[\theta,\theta](x,y)$, so that $d\alpha = -\frac{1}{2}\alpha[\theta,\theta]$. It is then clear that (5.74) and (5.76) are equivalent.

We turn now to the discussion of primitive elements for a reductive Lie algebra $\underline{\underline{g}}$. They are the elements of the space

$$P^{\cdot}_{\underline{\underline{g}}} = \sigma I^{\cdot}(\underline{\underline{g}})^{+} \subset (\wedge^{\cdot}\underline{\underline{g}}^*)^{\underline{\underline{g}}}$$

They are here not only given as cohomology classes, but in an explicit realization on the invariant cochain level. Note that for $\Phi,\Psi \in I^{+}(\underline{\underline{g}})$ and a $\underline{\underline{g}}$-invariant T we have

$$(5.76') \qquad \sigma(\Phi\Psi) = 0.$$

This follows from the immediately verified formula

$$d(T(\Phi\Psi) - \tfrac{1}{2}(T\Phi.\Psi + \Phi.T\Psi)) = 0.$$

The contractibility of $W(\underline{\underline{g}})^{\underline{\underline{g}}}$ implies namely

$$T(\Phi\Psi) = \tfrac{1}{2}(T\Phi.\Psi + \Phi.T\Psi) + d\chi$$

for some $\chi \in W(\underline{\underline{g}})^{\underline{\underline{g}}}$. Since $\pi\Phi = 0$, $\pi\Psi = 0$ and $\pi d\chi = d\pi\chi = 0 \in (\wedge\underline{\underline{g}}^*)^{\underline{\underline{g}}}$, the desired property (5.76') follows.

The suspension σ induces an isomorphism $I^{+}/(I^{+})^2 \xrightarrow{\cong} P_{\underline{\underline{g}}}$. A transgression $\tau = \tau_{\underline{\underline{g}}}$ is a linear map

$$\tau : P_{\underline{\underline{g}}} \longrightarrow I(\underline{\underline{g}})^{+}$$

splitting σ. (It is the transgression in the spectral sequence of $W(\underline{\underline{g}})^{\underline{\underline{g}}}$.) Note that for a choice of a transgression τ and a pair $y \in P_{\underline{\underline{g}}}$, $c = \tau y \in I(\underline{\underline{g}})^{+}$, a transgressive operator T defines an element $w = Tc \in W(\underline{\underline{g}})^{\underline{\underline{g}}}$ such that

$$dw = c \quad \text{and} \quad \pi(w) = \sigma c = y.$$

The element y is called transgressive, the element c the transgression of y, and w is called a transgressive cochain.

Let $y_1,\ldots,y_r$ be a basis of $P_{\underline{\underline{g}}}$, and $c_1,\ldots,c_r \in I(\underline{\underline{g}})$ transgressions $c_j = \tau y_j$. Then (compare [K 1], [GHV])

$$I(\underline{\underline{g}}) \cong K[c_1,\ldots,c_r]$$

and

$$H(\underline{\underline{g}}) \cong \wedge(y_1,\ldots,y_r).$$

The number r equals the rank of $\underline{\underline{g}}$ (the dimension of a Cartan subalgebra of $\underline{\underline{g}}$, i.e. a nilpotent subalgebra equal to its own normalizer).

5.77 A-COMPLEX [KT 5]. We define a complex realizing the cohomology $H(E_{\underline{\underline{h}}})$ for a commutative $\underline{\underline{g}}$-DG-algebra E with connections and where $(\underline{\underline{g}},\underline{\underline{h}})$ is a reductive pair of Lie algebras. Recall that this is a pair where $\underline{\underline{g}}$ is reductive and the adjoint representation of $\underline{\underline{h}}$ in $\underline{\underline{g}}$ semi-simple. $\underline{\underline{h}}$ is then also reductive. We assume that E satisfies the condition

(5.78) E^q is a direct sum of finite-dimensional simple $\underline{\underline{g}}$-modules for $q \geq 0$.

First we define the graded algebra

$$A^{\bullet}(E,\underline{\underline{h}}) = \wedge^{\bullet} P^{\bullet}_{\underline{\underline{g}}} \otimes E^{\bullet}_{\underline{\underline{g}}} \otimes I^{\bullet}(\underline{\underline{h}}), \tag{5.79}$$

where $P^{\bullet}_{\underline{\underline{g}}} \subset \wedge^{\bullet}(\underline{\underline{g}}^*)^{\underline{\underline{g}}}$ denotes the graded subspace of primitive elements of $\underline{\underline{g}}$. Let $\tau_{\underline{\underline{g}}} : P_{\underline{\underline{g}}} \longrightarrow I(\underline{\underline{g}})^+$ be a fixed transgression for $\underline{\underline{g}}$. A differential d_A is defined on A as a derivation of degree 1, which is zero on $I(\underline{\underline{h}})$, equal to the restriction of d_E on $E_{\underline{\underline{g}}}$, and on $\wedge P_{\underline{\underline{g}}}$ uniquely characterized by the formula

$$d_A(y) = 1 \otimes h(c) \otimes 1 - 1 \otimes 1 \otimes i^*(c) \text{ for } y \in P_{\underline{\underline{g}}} \text{ and } c = \tau_{\underline{\underline{g}}}(y). \tag{5.80}$$

The map $h : I(\underline{\underline{g}}) \longrightarrow E_{\underline{\underline{g}}}$ denotes the restriction of the

(Weil-homomorphism of the) connection $k : W(\underline{\underline{g}}) \longrightarrow E$ to $\underline{\underline{g}}$-basic elements, i.e. the Chern-Weil homomorphism of k. The map $i^* : I(\underline{\underline{g}}) \longrightarrow I(\underline{\underline{h}})$ denotes the canonical restriction. $d_A^2 = 0$ is trivially verified. The DG-algebra (A, d_A) is functorial with respect to connection preserving homomorphisms of $\underline{\underline{g}}$-DG-algebras and with respect to inclusions $\underline{\underline{h}}' \subset \underline{\underline{h}}$.

Next we define a homomorphism of DG-algebras

$$\text{(5.81)} \qquad \varphi(E,\underline{\underline{h}}) : A(E,\underline{\underline{h}}) \longrightarrow (E \otimes W(\underline{\underline{h}}))_{\underline{\underline{h}}}$$

which is natural in E and $\underline{\underline{h}}$. On $I(\underline{\underline{h}})$ this map is induced by the canonical map $W(\underline{\underline{h}}) \longrightarrow E \otimes W(\underline{\underline{h}})$. On $E_{\underline{\underline{g}}}$ the map is induced by the canonical map $E \longrightarrow E \otimes W(\underline{\underline{h}})$. On $\wedge P_{\underline{\underline{g}}}$ the map is the canonical extension of $-\lambda^1(E,\underline{\underline{h}}) \circ \tau_{\underline{\underline{g}}}$, where $\lambda^1(E,\underline{\underline{h}}) = (k \otimes W(i)) \circ \lambda^1$. Here $W(i)$ is induced by $i : \underline{\underline{h}} \subset \underline{\underline{g}}$ and $\lambda^1 : W(\underline{\underline{g}}) \longrightarrow W(\underline{\underline{g}}) \otimes W(\underline{\underline{g}})$ is the universal homotopy operator of 5.54. Note that for $E = W(\underline{\underline{g}})$ and $\underline{\underline{h}} = 0$ the map $-\lambda^1(E,\underline{\underline{h}}) = -\lambda^1(W(\underline{\underline{g}}),0)$ is the universal transgressive operator T_o of (5.70).

For any $\underline{\underline{h}}$-equivariant splitting $\theta : \underline{\underline{g}} \longrightarrow \underline{\underline{h}}$ of the exact sequence $0 \longrightarrow \underline{\underline{h}} \longrightarrow \underline{\underline{g}} \longrightarrow \underline{\underline{g}}/\underline{\underline{h}} \longrightarrow 0$ the map $\alpha = (\mathrm{id}, k \circ k(\theta)) : E \otimes W(\underline{\underline{h}}) \longrightarrow$ induces by (5.62) a homology isomorphism

$$\alpha_* : H(E \otimes W(\underline{\underline{h}}))_{\underline{\underline{h}}}) \xrightarrow{\cong} H(E_{\underline{\underline{h}}}).$$

Note that $\alpha \circ \lambda^1(E,\underline{\underline{h}}) = k \circ \lambda^1(\theta)$ in the notation of proposition 5.67.

The composition of the natural transformation $\varphi(E,\underline{\underline{h}})$ of (5.81) with the homology equivalence α leads to the following result.

5.82 THEOREM [KT 5]. <u>The homomorphism</u> (5.81) <u>induces an isomorphism</u>, and together with 5.62 isomorphisms

$$H(A(E,\underline{\underline{h}})) \xrightarrow[\cong]{\varphi(E,\underline{\underline{h}})_*} H((E \otimes W(\underline{\underline{h}}))_{\underline{\underline{h}}}) \xrightarrow[\cong]{\alpha_*} H(E_{\underline{\underline{h}}}).$$

For the proof we need the assumption that the pair $(\underline{\underline{g}},\underline{\underline{h}})$ is reductive. For $\underline{\underline{h}} = 0$ this is a result of Chevalley. This Theorem is proved by introducing filtrations on $A(E,\underline{\underline{h}})$ and $(E \otimes W(\underline{\underline{h}}))_{\underline{\underline{h}}}$ which are preserved by $\varphi(E,\underline{\underline{h}})$, and establishing that φ induces an isomorphism of the initial terms of the associate spectral sequences. The following two multiplicative filtrations are used.

First the canonical filtration on $W(\underline{\underline{h}})$ induces a filtration on $I(\underline{\underline{h}})$ and hence on $A(E,\underline{\underline{h}})$ via $I(\underline{\underline{h}})$, and further on $E \otimes W(\underline{\underline{h}})$ and hence on $(E \otimes W(\underline{\underline{h}}))_{\underline{\underline{h}}}$ via $W(\underline{\underline{h}})$. These even filtrations will be denoted $F^{2p}(\underline{\underline{h}}) = {}'F^{2p}$ and are called $\underline{\underline{h}}$-filtrations of the respective DG-algebras.

Next consider the canonical filtration on E given by 5.14. It induces a filtration on $E_{\underline{\underline{g}}}$ and hence a (decreasing) filtration on $A(E,\underline{\underline{h}})$, and further on $E \otimes W(\underline{\underline{h}})$ and hence on $(E \otimes W(\underline{\underline{h}}))_{\underline{\underline{h}}}$ via E. These filtrations will be denoted by $F^p(\underline{\underline{g}}) = {}''F^p$ and are called $\underline{\underline{g}}$-filtrations of the respective DG-algebras.

The natural homomorphism φ is filtration preserving for both the $\underline{\underline{h}}$- and the $\underline{\underline{g}}$-filtrations.

5.83 THEOREM.

(i) φ <u>induces for the even spectral sequences associated to the $\underline{\underline{h}}$-filtrations an isomorphism on the</u> ${}'E_{2r}$<u>-level for</u> $r \geq 1$.

(ii) <u>The composition of</u> φ <u>with the canonical map</u> $(E \otimes W(\underline{\underline{h}}))_{\underline{\underline{h}}} \xrightarrow{\alpha} E_{\underline{\underline{h}}}$ <u>is filtration preserving with respect to the $\underline{\underline{g}}$-filtration on</u> $A(E,\underline{\underline{h}})$ <u>and</u> $E_{\underline{\underline{h}}}$. <u>It induces for the associated spectral sequences an isomorphism on the</u> ${}''E_r$<u>-level for</u> $r \geq 1$.

Theorem 5.82 is now a consequence of 5.83, (i) and 5.62. Together with the computation of the initial terms of the spectral sequences associated to the $\underline{\underline{h}}$- and $\underline{\underline{g}}$-filtrations on $A(E,\underline{\underline{h}})$ we obtain the following result.

5.84 THEOREM

(i) <u>There is a multiplicative even spectral sequence</u>

$$'E_2^{2p,q} = H^q(E) \otimes I^{2p}(\underline{\underline{h}}) \Longrightarrow H^{2p+q}(E_{\underline{\underline{h}}}).$$

(ii) <u>There is a multiplicative spectral sequence</u>

$$''E_2^{p,q} = H^q(\underline{\underline{g}},\underline{\underline{h}}) \otimes H^p(E_{\underline{\underline{g}}}) \Longrightarrow H^{p+q}(E_{\underline{\underline{h}}}).$$

Some of the preceding results have independently been obtained by Halperin under slightly less restrictive hypotheses on E (see [GHV], vol. III).

A geometric analogon of the previous results concerns the G-DG-algebra $\Omega(P)$ for an ordinary principal G-bundle $P \longrightarrow M$ with compact group G. For a closed subgroup $H \subset G$ the natural homomorphisms

$$A(\Omega(P),H) \longrightarrow (\Omega(P) \otimes W(\underline{\underline{h}}))_H \longrightarrow \Omega(P/H)$$

induce isomorphisms in homology. The spectral sequences discussed above have as geometric analoga the Serre spectral sequences of the fibrations

$$P \longrightarrow E_H \times_H P \xrightarrow{\pi'} B_H, \quad G/H \longrightarrow P/H \xrightarrow{\pi''} M$$

where $E_H \longrightarrow B_H$ denotes a universal H-bundle.

These theorems apply to $E = W(\underline{\underline{g}})_k$, $k \geq 0$. It is convenient to set $W(\underline{\underline{g}})_\infty = W(\underline{\underline{g}})$. Then we obtain the following result.

5.85 THEOREM. For $0 \leq k \leq \infty$ the cohomology $H(W(\underline{\underline{g}},\underline{\underline{h}})_k)$ can be computed as the cohomology of the DG-algebra

$$(5.86)\qquad A(W(\underline{\underline{g}})_k,\underline{\underline{h}}) = \wedge^{\cdot} P^{\cdot}_{\underline{\underline{g}}} \otimes I^{\cdot}(\underline{\underline{g}})_k \otimes I^{\cdot}(\underline{\underline{h}})$$

There are multiplicative spectral sequences

$$(5.87)\qquad {}'E_2^{2p,q} = H^q(W(\underline{\underline{g}})_k) \otimes I^{2p}(\underline{\underline{h}}) \Longrightarrow H^{2p+q}(W(\underline{\underline{g}},\underline{\underline{h}})_k)$$

$$(5.88)\qquad {}''E_2^{2p,q} = H^q(\underline{\underline{g}},\underline{\underline{h}}) \otimes I^{2p}(\underline{\underline{g}})_k \Longrightarrow H^{2p+q}(W(\underline{\underline{g}},\underline{\underline{h}})_k).$$

In the second spectral sequence $''E_2^{2p+1,q} = 0$.

For $k = \infty$ we have $'E_2^{2p,q} = 0$ for $q > 0$ and an edge isomorphism $I(\underline{\underline{h}}) \xrightarrow{\cong} H(W(\underline{\underline{g}},\underline{\underline{h}}))$. Therefore

$$(5.89)\qquad {}''E_2^{2p,q} = H^q(\underline{\underline{g}},\underline{\underline{h}}) \otimes I^{2p}(\underline{\underline{g}}) \longrightarrow I^{2p+q}(\underline{\underline{h}})$$

where the edge homomorphism $I^{2p}(\underline{\underline{g}}) \longrightarrow I^{2p}(\underline{\underline{h}})$ is the restriction map. Since $I(\underline{\underline{g}}) = S(\underline{\underline{g}}^*)^{\underline{\underline{g}}}$, the initial term equals $H^q(\underline{\underline{g}},\underline{\underline{h}};S^p(\underline{\underline{g}}^*))$.

For the case of a connected group G and maximal compact subgroup K this gives e.g. a spectral sequence

$$(5.90)\qquad H^q(\underline{\underline{g}},\underline{\underline{k}}\ ;\ S^p(\underline{\underline{g}}^*)) \Longrightarrow I(\underline{\underline{k}}).$$

The initial term can by the Van Est Theorem [E] be replaced by the continuous cohomology $H_c(G,S^p(\underline{\underline{g}}^*))$, whereas the end term is by the universal Chern-Weil homomorphism isomorphic to $H(B_K) \cong H(B_G)$. Under these replacements (5.91) coincides with the spectral sequence

$$H_c^q(G,S^p(\underline{\underline{g}}^*)) \Longrightarrow H(B_G)$$

considered in [B 5] [SH 1].

5.91 DIFFERENCE CONSTRUCTION FOR Δ_*. In this section we return to a foliated G-bundle with H-reduction and evaluate the generalized characteristic homomorphism Δ_* on the A-complex realizing $H(W(\mathfrak{g},H)_q)$. To be able to apply the previous results proved in a purely Lie algebra context, we assume here as in all later geometric applications that G is either connected or $I(G) \cong I(G_0) \equiv I(\mathfrak{g})$, and H has finitely many components.

With the notations of chapter 4 and 5 the problem is to determine the composition

$$(5.92)\quad \tilde{\Delta}(\omega) : A(W(\mathfrak{g})_q, H) \xrightarrow{\varphi} (W(\mathfrak{g})_q \otimes W(\mathfrak{h}))_H \xrightarrow{\alpha} W(\mathfrak{g},H)_q \xrightarrow{\Delta(\omega)} \Omega(M)$$

We recall the definition of the ocurring maps. The map $\varphi = \varphi(W(\mathfrak{g})_q, H)$ is the homology equivalence (5.81). The map $\alpha = (\mathrm{id}, k(\theta))$ is defined by the Weil homomorphism $k(\theta): W(\mathfrak{h}) \longrightarrow W(\mathfrak{g})$ of the connection $\mu \circ \wedge\theta^* : \wedge\mathfrak{h}^* \longrightarrow \wedge\mathfrak{g}^* \longrightarrow W(\mathfrak{g})$ given by an H-equivariant splitting $\theta : \mathfrak{g} \longrightarrow \mathfrak{h}$ of the exact H-module sequence $0 \longrightarrow \mathfrak{h} \longrightarrow \mathfrak{g} \longrightarrow \mathfrak{g}/\mathfrak{h} \longrightarrow 0$. $k(\theta)$ is an H-DG-homomorphism and α a homology equivalence by 5.62. The map $\Delta(\omega)$ is the generalized characteristic homomorphism on the cochain level. By construction $\Delta(\omega) = s^* \circ k(\omega)_H$, where $s : M \longrightarrow P/H$ defines the H-reduction of P, and $k(\omega)_H$ is the Weil-homomorphism of a connection ω adapted to the foliation of the bundle P.

Consider the diagram

$$(5.93)\quad \begin{array}{ccc} W(\mathfrak{g},H) & & \\ \uparrow{\scriptstyle\alpha} & \searrow^{k(\omega)_H} & \\ (W(\mathfrak{g}) \otimes W(\mathfrak{h}))_H & \xrightarrow{(k(\omega),\,k(\bar{\omega}'))_H} & \Omega(P/H) \end{array}$$

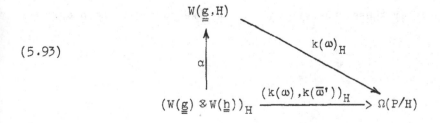

To explain $k(\overline{\omega}')$, let first ω' be any connection in the H-bundle $P' \longrightarrow M$. This bundle is defined as the pull-back under the section $s : M \longrightarrow P/H$ in the commutative diagram

$$\begin{array}{ccc} P' & \xrightarrow{\overline{s}} & P \\ \downarrow & & \downarrow \\ M & \underset{\hat{\pi}}{\overset{s}{\rightleftarrows}} & P/H \end{array}$$

It follows that there is a unique connection $\overline{\omega}'$ on the H-bundle $P \longrightarrow P/H$ such that $\overline{s}^*\overline{\omega}' = \omega'$. To show that with this interpretation of $\overline{\omega}'$ diagram (5.93) is homotopy commutative, we consider secondly the interpretation of $\overline{\omega}' = \theta \circ \omega$ as the connection defined in the H-bundle $P \longrightarrow P/H$ by ω and the split θ. In this situation we define ω' in P' by the formula $\omega' = \overline{s}^*\overline{\omega}'$. With this second interpretation of $\overline{\omega}'$ the diagram (5.93) is even strictly commutative, since clearly

$$k(\overline{\omega}') = k(\omega) \circ k(\theta) : W(\underline{\underline{h}}) \longrightarrow \Omega(P).$$

But the two constructions for $k(\overline{\omega}')$ just discussed are by the construction of 5.58 homotopic by a homotopy preserving the operators $i(x)$ and $\theta(x)$ for $x \in \underline{\underline{h}}$. Therefore diagram (5.93) is commutative in homology for the first interpretation of $k(\overline{\omega}')$. Since the critical filtration preserving property concerns only the map $k(\omega)$ which is the same in both interpretations, it follows that the diagram corresponding to (5.93) after replacement of $W(\underline{\underline{g}})$ by $W(\underline{\underline{g}})_q$ induces still a commutative diagram in homology.

After these observations on $\overline{\omega}'$ we consider the diagram

$$
(5.94)\qquad
\begin{array}{ccc}
W(\mathfrak{g},H)_q & & \\
\uparrow{\scriptstyle\alpha} & \searrow{\scriptstyle k(\omega)_H} & \\
(W(\mathfrak{g})_q \otimes W(\mathfrak{h}))_H & \xrightarrow{\ (k(\omega),k(\overline{\omega}'))_H\ } & \Omega(P/H) \\
\uparrow{\scriptstyle\varphi} & & \downarrow{\scriptstyle s^*} \\
A(W(\mathfrak{g})_q,H) & \xrightarrow{\ \tilde{\Delta}(\omega)\ } & \Omega(M)
\end{array}
$$

It is commutative with the second interpretation of $\overline{\omega}'$, and commutative in homology with the first interpretation of $\overline{\omega}'$. In any case we can now evaluate $\tilde{\Delta}(\omega)$ via the (strictly) commutative rectangle in (5.94).

5.95 THEOREM. <u>With the preceding notations, the evaluation</u> $\tilde{\Delta}(\omega)$ <u>on the generalized characteristic homomorphism</u> Δ_* <u>of a foliated G-bundle</u> P <u>on the complex</u>

$$A(W(\mathfrak{g})_q,H) = \wedge P_{\mathfrak{g}} \otimes I(G)_q \otimes I(H)$$

<u>is as follows</u>

(i) $\tilde{\Delta}(\omega)\Phi = h(\omega)\Phi$ <u>for</u> $\Phi \in I(G)_q$

(ii) $\tilde{\Delta}(\omega)\Psi = h(\omega')\Psi$ <u>for</u> $\Psi \in I(H)$

(iii) $\tilde{\Delta}(\omega)x = -\lambda^1(\omega,\omega')\Phi$ <u>for</u> $x \in P_{\mathfrak{g}}$ <u>and</u> $\Phi = \tau x \in I(G)^+$.

In this statement h and h' denote the characteristic homomorphism of the connections ω and ω'. The map $\lambda'(\omega,\omega')$ is the composition

$$(5.96)\quad I(G) \xrightarrow{\lambda^1} (W(\mathfrak{g})\otimes W(\mathfrak{g}))_G \xrightarrow{id\otimes W(i)} (W(\mathfrak{g})\otimes W(\mathfrak{h}))_H \xrightarrow{(k(\omega),\,k(\overline{\omega}'))_H} \Omega(P/H) \rightarrow$$

$$\xrightarrow{s^*} \Omega(M)$$

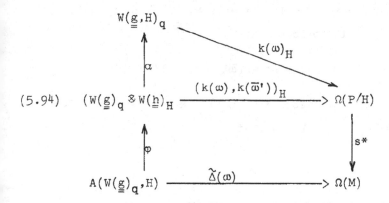

where λ^1 is induced in G-basic elements by λ^1 in 5.54. By (i) in 5.54 and 5.96 it follows that

$$d\lambda^1(\omega,\omega') + \lambda^1(\omega,\omega')d = s^* \circ h(\overline{\omega}') \circ i^* - h(\omega)$$

Note that $\overline{s}^*\overline{\omega}' = \omega'$ implies $\overline{s}^* \circ k(\overline{\omega}\,') = k(\omega')$ and therefore $\overline{s}^* \circ k(\overline{\omega}')_H = k(\omega')_H$. This shows that

$$s^* \circ h(\overline{\omega}') = h(\omega').$$

Since further $d\Phi = 0$ for $\Phi \in I(G)$, it follows that

$$(5.97) \qquad d\lambda^1(\omega,\omega')\Phi = h(\omega')i^*\Phi - h(\omega)\Phi \quad \text{for} \quad \Phi \in I(G)^+$$

The verification of the formulas in theorem 5.95 is now as follows. By definition of φ and (5.94) for $\Phi \in I(G)_q$

$$\tilde{\Delta}(\omega)\Phi = \tilde{\Delta}(\omega)\ (1 \otimes \Phi \otimes 1) = s^* \circ k(\omega)_H)\Phi = h(\omega)\Phi$$

since $k(\omega)\Phi$ is H-basic in $\Omega(P)$ (in fact G-basic). For $\Psi \in I(H)$

$$\tilde{\Delta}(\omega)\Psi = \tilde{\Delta}(\omega)\ (1 \otimes 1 \otimes \Psi) = s^* \circ k(\overline{\omega}\,')_H)\Psi = h(\omega')\Psi.$$

For $x \in P_{\underline{\underline{g}}}$ and $\Phi = \tau x$ by definition

$$\varphi(W(\underline{\underline{g}})_q, H)x = -\ (id \otimes W(i) \circ \lambda^1(\Phi)$$

so that by (5.94) (5.96)

$$\tilde{\Delta}(\omega)x = -s^* \circ (k(\omega), k(\overline{\omega}'))_H \circ (id \otimes W(i)) \circ \lambda\ (\Phi)$$

$$= -\lambda^1(\omega,\omega')\Phi$$

This finishes the verification of the evaluation formulas in theorem 5.95.

It is of interest to consider the case of $\Phi \in I^{2p}(G)$

such that $i^*\Phi = 0 \in I(H)$. By (5.97) we have then

$$d\lambda^1(\omega,\omega')\Phi = -h(\omega)\Phi$$

For the suspension $x = \sigma\Phi$ the element $x \otimes 1 \otimes 1 \in A(W(\underline{\underline{g}})_q, H)$ is a cycle precisely if $h(\omega)\Phi = 0$, as follows from 5.80. If this is the case, then

$$\tilde{\Delta}(\omega)x = -\lambda^1(\omega,\omega')\Phi \in \Omega^{2p-1}(M)$$

is a closed form defining a secondary characteristic class.

Constructions of this type occur in the work of Chern-Simons [CS 1], and it was one of our motivations for the definition of the A-complex.

If we return for a moment to the notations in theorem 4.52 and its proof, the definition of the subcomplex K_q in (4.57) is given by

$$K_q = \ker(A \longrightarrow I(G)_q \otimes_{I(G)} I(H)).$$

Clearly

$$B = \Lambda^+P \otimes I(G)_q \otimes I(H) = \ker(A \longrightarrow I(G)_q \otimes I(H)).$$

Denote

$$J = \mathrm{Id}\{h(\Phi) \otimes 1 - 1 \otimes i^*\Phi\} \subset I(G)_q \otimes I(H)$$

the ideal generated in the RHS by the elements $\{\ \}$ for $\Phi \in I(G)^+$. Then it follows easily that

$$K_q = B + A \,.\, 1 \otimes J = B \oplus (1 \otimes J)$$

In fact since by (5.80) clearly $1 \otimes J \subset dA$, $H(K_q)$ is generated by cocycles in B (but $B \subset K_q$ is not a subcomplex by (5.80)).

For $q = 0$ we have in particular

$$K_0 = \wedge^+ P \otimes I(H) \oplus 1 \otimes I(G)^+ . I(H)$$

for the complex occurring in theorem 4.52.

5.98 $\hat{A}$-COMPLEX [KT 5]. The computation of $H(A(E,\underline{\underline{h}}))$ can be further simplified. We return to a purely Lie algebraic context. We need the Samelson space $\hat{P} \subset P_{\underline{\underline{g}}}$ of a reductive pair $(\underline{\underline{g}},\underline{\underline{h}})$ [CA] [GHV]. It is defined via the cohomology map $\gamma_*: H(\underline{\underline{g}},H) \to H(\underline{\underline{g}})$ induced by the inclusion $\gamma: (\wedge \underline{\underline{g}}^*)_H \longrightarrow \wedge \underline{\underline{g}}^*$ and the Hopf-Samelson isomorphism $\varphi : \wedge P_{\underline{\underline{g}}} \xrightarrow{\cong} H(\underline{\underline{g}})$ as

(5.99) $$\hat{P} = P_{\underline{\underline{g}}} \cap \operatorname{im}(\varphi^{-1} \circ \gamma_*)$$

We make the following assumption on $(\underline{\underline{g}},\underline{\underline{h}})$:

(5.100) $$C : \dim \hat{P} = \dim \underline{\underline{g}} - \dim \underline{\underline{h}} \quad \text{(Cartan pair)},$$

and even the stronger assumption

(5.101) CS : there exists a trangression τ for $\underline{\underline{g}}$ such that

$$\ker(I(\underline{\underline{g}}) \longrightarrow I(\underline{\underline{h}})) = \operatorname{Id}(\tau\hat{P}) \subset I(\underline{\underline{g}}) \quad \text{(special Cartan pair)}.$$

This condition is satisfied for symmetric pairs $(\underline{\underline{g}},\underline{\underline{h}})$ and for pairs $(\underline{\underline{g}},\underline{\underline{h}})$ of equal rank. (CS) implies (C).

The point of the concept (CS) introduced in [KT 5] is that then

(5.102) $$\hat{P} = \sigma_{\underline{\underline{g}}} \ker(I(\underline{\underline{g}}) \longrightarrow I(\underline{\underline{h}}))$$

which simplifies the determination of the Samelson space. In fact in all our applications the pairs $(\underline{\underline{g}},\underline{\underline{h}})$ are (CS)-pairs.

The inclusion $\sigma_{\underline{\underline{g}}} \ker(I(\underline{\underline{g}}) \longrightarrow I(\underline{\underline{h}})) \subset \hat{P}$ is a consequence of the following result.

5.103 PROPOSITION. <u>Let</u> $\Phi \in \ker(I(\mathfrak{g}) \longrightarrow I(\mathfrak{h}))$. <u>Then</u> $\sigma_{\mathfrak{g}}\Phi \in \Lambda\mathfrak{g}^*$ <u>is</u> $\mathfrak{h}$<u>-basic</u>.

Proof. This is based on the homology equivalence $k(\theta)_{\mathfrak{h}} : I(\mathfrak{h}) \longrightarrow W(\mathfrak{g},\mathfrak{h})$ defined in (5.65) by an $\mathfrak{h}$-equivalent splitting $\theta : \mathfrak{g} \longrightarrow \mathfrak{h}$ of the exact $\mathfrak{h}$-module sequence $0 \longrightarrow \mathfrak{h} \longrightarrow \mathfrak{g} \longrightarrow \mathfrak{g}/\mathfrak{h} \longrightarrow 0$.

Consider the diagram

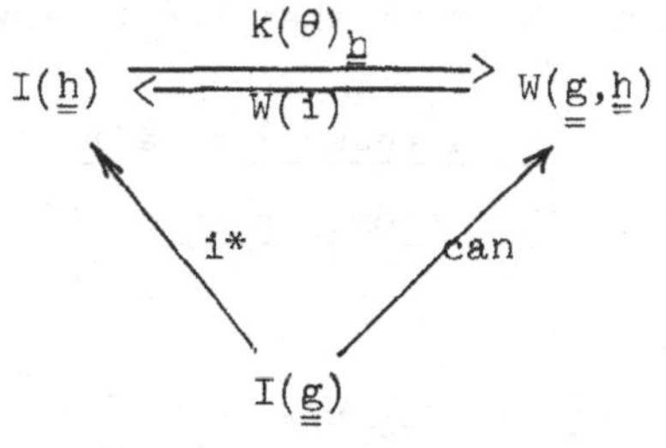

where can denotes the inclusion $I(\mathfrak{g}) = W(\mathfrak{g})_{\mathfrak{g}} \subset W(\mathfrak{g})_{\mathfrak{h}} = W(\mathfrak{g},\mathfrak{h})$. Since $W(i) \circ \mathrm{can} = i^*$, and $k(\theta)_{\mathfrak{h}}$ induces a homology inverse of $W(i)$ by theorem (5.64), it follows that $k(\theta)_{\mathfrak{h}} \circ i^*$ and can have the same effect in homology.

Let $\Phi \in \ker(I(\mathfrak{g}) \longrightarrow I(\mathfrak{h}))$. By the preceding $\Phi = d\Psi$ in $W(\mathfrak{g},\mathfrak{h})$ and therefore Ψ is a transgressive cochain of Φ. The canonical projection $\pi : W(\mathfrak{g}) \longrightarrow \Lambda\mathfrak{g}^*$ maps $W(\mathfrak{g},\mathfrak{h}) \longrightarrow (\Lambda\mathfrak{g}^*)_{\mathfrak{h}}$ and $\sigma\Phi = \pi\Psi$ (see 5.73). Therefore $\sigma\Phi$ is $\mathfrak{h}$-basic. □

We define now for a commutative $\mathfrak{g}$-DG-algebra E with connection and satisfying 5.78 a graded algebra

$$\hat{A}^{\cdot}(E) = \Lambda^{\cdot}\hat{P}^{\cdot} \otimes E_{\mathfrak{g}} \tag{5.104}$$

with differential characterized by $d_{\hat{A}}(y) = 1 \otimes h(c)$ for $y \in \hat{P}$, $c = \tau y \in I(\mathfrak{g})$, and equal to the restriction of d_E on $E_{\mathfrak{g}}$. Under the condition 5.101 for the pair $(\mathfrak{g},\mathfrak{h})$ we may choose the fixed transgression $\tau = \tau_{\mathfrak{g}}$ as in (CS), and then the canonical map

$$(5.105) \qquad \hat{A}(E) \xrightarrow{\ i\ } A(E,\underline{\underline{h}})$$

given by $x \otimes c \longrightarrow x \otimes c \otimes 1$ is compatible with the differentials. Consider further the canonical homomorphism $j : I(\underline{\underline{h}}) \longrightarrow A(E,\underline{\underline{h}})$ and the induced cohomology map

$$(5.106) \qquad \beta = (i_*, j_*) : H(\hat{A}(E)) \otimes I(\underline{\underline{h}}) \longrightarrow H(A(E,\underline{\underline{h}})).$$

Note that $\hat{A}(E)$ and hence $H(\hat{A}(E))$ is an $I(\underline{\underline{g}})$-module via the characteristic homomorphism. We have then the following result.

(5.107) THEOREM. Let $(\underline{\underline{g}},\underline{\underline{h}})$ be a reductive pair satisfying the condition (CS) in (5.101). The homomorphism 5.106 factorizes through an isomorphism

$$\bar{\beta} : H(\hat{A}(E)) \otimes_{I(\underline{\underline{g}})} I(\underline{\underline{h}}) \xrightarrow{\ \cong\ } H(A(E,\underline{\underline{h}})).$$

This result applies to $E = W(\underline{\underline{g}})_k$. Let $\hat{A}_k = \hat{A}(W(\underline{\underline{g}})_k)$. Then we get the following consequence

5.108 THEOREM. For $0 \leq k \leq \infty$ there are isomorphisms

$$H(\hat{A}_k) \otimes_{I(\underline{\underline{g}})} I(\underline{\underline{h}}) \xrightarrow{\ \cong\ } H(A(W(\underline{\underline{g}})_k,\underline{\underline{h}}) \cong H(W(\underline{\underline{g}},\underline{\underline{h}})_k).$$

This is a purely algebraic result, valid over any groundfield K of characteristic zero. To describe a basis of $H(\hat{A}_k)$ over K, let $I(\underline{\underline{g}}) \cong K[c_1,\ldots,c_r]$, $r = \operatorname{rank} \underline{\underline{g}}$ (ordered such that $\deg c_i \leq \deg c_{i+1}$). Then $P_{\underline{\underline{g}}}$ has a basis of elements transgressing to $c_1,\ldots,c_r$ respectively. Let $r' = \operatorname{rank} \underline{\underline{g}} - \operatorname{rank} \underline{\underline{h}} = \dim \hat{P}$ and $y_1,\ldots,y_{r'}$ a basis of $\hat{P}$ such that y_ℓ transgresses to c_{α_ℓ} $(\alpha_1 \leq \cdots \leq \alpha_{r'})$. For $\underline{\underline{h}} = 0$ we have in particular $\hat{P} = P_{\underline{\underline{g}}}$, $r' = r$ and $\alpha_\ell = \ell$ for all ℓ. With these notations

$$\hat{A}_k = \Lambda(y_1,\ldots,y_{r'}) \otimes K[c_1,\ldots,c_r]_k$$

where $dy_\ell = c_{\alpha_\ell}$. We use the following conventions:

(5.109)
$$y_{(i)} = y_{i_1}\wedge\ldots\wedge y_{i_s} \text{ for } (i) = (i_1,\ldots,i_s),\ 1 \leq i_1 < \ldots < i_s \leq r' \ (s>0)$$
$$y_{(i)} = 1 \text{ for } (i) = \phi \ (s=0);$$
$$c_{(j)} = c_1^{j_1}\ldots c_r^{j_r} \text{ for } (j) = (j_1,\ldots,j_r),\ 0 \leq j_i;$$
$$2p = \deg c_{(j)} = \sum_{i=1}^{r} j_i \deg c_i.$$

Then we have the following result (this is the result of [KT 5] with a slight change of notation adapted to the present purposes).

5.110 THEOREM. A K-basis of $H(\hat{A}_k)$ is given by the classes of the monomial cocycles $z_{(i,j)} = y_{(i)} \otimes c_{(j)}$ satisfying the conditions:

(a) $0 \leq 2p \leq 2k$, $0 \leq s \leq r'$;

(b) $\deg c_{\alpha_{i_1}} = \deg y_{i_1} + 1 \geq 2(k+1-p)$ if $(i) \neq \phi$;

(c) $j_{\alpha_\ell} = 0$ for $\ell < i_1$ if $(i) \neq \phi$ and $j_{\alpha_\ell} = 0$ for all ℓ if $(i) = \phi$.

Remarks. (i) The monomials $z_{(\phi,j)}$ $(s=0)$ form a basis of the primary classes (induced from $I(\underline{\underline{g}})_k \longrightarrow W(\underline{\underline{g}},\underline{\underline{h}})_k$). (ii) The monomials $z_{(i,j)}$ for $s > 0$ form a basis for the secondary classes. (iii) The classes for $s > 0$ and $p = 0$ are the classes $y_{(i)} \otimes 1$ with $\deg y_{i_1} + 1 \geq 2(k+1)$.

Note that the degrees of the secondary classes $[z_{(i,j)}]$ in $H(\hat{A}_k)$ satisfy the inequality

(5.111)
$$2k+1 \leq \deg z_{(i,j)} \leq 2k+m,\ m = \dim \underline{\underline{g}}.$$

In fact $s > 0$ guarantees the occurance of at least the element

y_{i_1} and hence

$$\deg z_{(i,j)} + 1 \geqq 2p + \deg y_{i_1} + 1 \geqq 2(k+1).$$

The other inequality follows from the fact that $\deg c_{(j)} \leqq 2k$ and $\deg y_{(i)} \leqq m$ (which equals the sum of the degrees of all primitive generators).

5.112 REMARK ON DIFFERENCE CONSTRUCTION. We return to the geometric context of 5.91. With the construction of the subcomplex $\hat{A} \subset A$ it is natural to consider the composition

$$(5.113) \qquad \hat{\Delta}(\omega) : \hat{A}(W(\underline{\underline{g}})_q) \longrightarrow A(W(\underline{\underline{g}})_q, H) \xrightarrow{\tilde{\Delta}(\omega)} \Omega(M)$$

The evaluation of $\hat{\Delta}(\omega)$ is then clearly also given by (5.95), (i) and (iii). The characteristic homomorphism $\Delta_* \sim \tilde{\Delta}_*$ is realized by

$$(5.114) \qquad \hat{\Delta}_* \otimes h'_* : H(\hat{A}(W(\underline{\underline{g}})_q) \otimes_{I(G)} I(H) \longrightarrow H_{DR}(M)$$

where h'_* is the characteristic homomorphism of the H-reduction P' of P. As mentioned in the outline of this chapter, this evaluation $\hat{\Delta}_*$ of the characteristic homomorphism is similar to the construction of a characteristic homomorphism in [B 3].

6. NON-TRIVIAL CHARACTERISTIC CLASSES FOR FLAT BUNDLES

In this chapter we give examples of flat bundles with non-trivial characteristic invariants.

The type of examples from [KT 1] discussed in theorem 4.87 and the rest of chapter 4 was the starting point of the authors' work presented in these lectures.

The relative Lie algebra cohomology $H(\underline{\underline{g}},H)$ of the pair (G,H) appears in the construction of the characteristic homomorphism Δ_* as the universal characteristic cohomology for flat G-bundles with an H-reduction. We assume throughout that G is either connected or $I(G) \cong I(G_0) \equiv I(\underline{\underline{g}})$ for the connected component G_0 of G (as e.g. for $GL(n,\mathbb{R})$ or $GL(n,\mathbb{C})$). The closed subgroup $H \subset G$ is assumed to have finitely many connected components. The reductivity of the pair (G,H) means that the pair $(\underline{\underline{g}},\underline{\underline{h}})$ is reductive. The exact sequence $0 \longrightarrow \underline{\underline{h}} \longrightarrow \underline{\underline{g}} \longrightarrow \underline{\underline{g}}/\underline{\underline{h}} \longrightarrow 0$ has then an H-equivariant splitting $\theta : \underline{\underline{g}} \longrightarrow \underline{\underline{h}}$, as follows by averaging an $\underline{\underline{h}}$-equivariant splitting over the group of components of H. With these assumptions on (G,H) the results of chapter 5 can be formulated in Lie group terms, and these assumptions are satisfied in the applications of main interest.

To make the applications in this chapter as easily accessible as possible, we review the concepts and results of chapter 5 to the extent needed for the computation of $H(\underline{\underline{g}},H)$.

The suspension map $\sigma : I(\underline{\underline{g}})^+ \longrightarrow H(\underline{\underline{g}})$ can be realized on the cochain level by the map $\sigma : I(\underline{\underline{g}})^+ \longrightarrow (\Lambda\underline{\underline{g}}^*)^{\underline{\underline{g}}}$ given by the formula (5.74). Its image determines the subspace of primitive elements $P_{\underline{\underline{g}}} \subset (\Lambda\underline{\underline{g}}^*)^{\underline{\underline{g}}}$ of $\underline{\underline{g}}$. The complex (5.79) realizing $H(\underline{\underline{g}},H)$ according to theorem 5.82 reduces to

(6.1) $$A(\wedge \underline{\underline{g}}^*, H) = \wedge P_{\underline{\underline{g}}} \otimes I(H).$$

The differential d_A is zero on $I(H)$, and as a derivation it is determined on elements of $P_{\underline{\underline{g}}}$. For an element $y \in P_{\underline{\underline{g}}}$ transgressing to $c \in I(G)$, we have by (5.80)

(6.2) $$d_A(y \otimes 1) = -1 \otimes i^*c ,$$

where $i^* : I(G) \longrightarrow I(H)$ is the restriction induced by $i : H \subset G$.

For the case of the trivial group $H = \{e\}$ and $A(\wedge \underline{\underline{g}}^*) = \wedge P_{\underline{\underline{g}}}$ theorem 5.82 reduces to the isomorphism of Hopf-Samelson

(6.3) $$\varphi : \wedge^{\cdot} P^{\cdot}_{\underline{\underline{g}}} \xrightarrow{\cong} H^{\cdot}(\underline{\underline{g}})$$

which holds for reductive Lie algebras [K 1]. In fact φ is induced by the inclusion $P_{\underline{\underline{g}}} \subset (\wedge \underline{\underline{g}}^*)^{\underline{\underline{g}}} \cong H(\underline{\underline{g}})$, which extends to a homomorphism $\wedge P_{\underline{\underline{g}}} \longrightarrow (\wedge \underline{\underline{g}}^*)^{\underline{\underline{g}}}$, since every element of $P_{\underline{\underline{g}}}$ is of odd degree and hence of square zero in $H(\underline{\underline{g}})$.

To simplify the determination of the Samelson space $\hat{P} \subset P_{\underline{\underline{g}}}$ of the part (G,H), we assume throughout that (G,H) is a special Cartan pair (CS-pair) in the sense of definition (5.101). Then

(6.5) $$\hat{P} = \sigma_{\underline{\underline{g}}} \ker(I(G) \longrightarrow I(H)).$$

In all our applications this assumption is satisfied, which makes the determination of $\hat{P}$ a simple matter.

By theorem (5.107) we have the isomorphism

(6.6) $$H(\underline{\underline{g}}, H) \cong \wedge \hat{P} \otimes I(H)/I(G)^{+}.I(H)$$

which is well-known [CA] [GHV, vol III].

The generalized characteristic homomorphism Δ_* of a G-flat bundle restricted to $I(H)|I(G)^+.I(H)$ is by theorem 4.52, (i) induced by the Chern-Weil homomorphism of the given H-reduction. Thus the invariants of interest are the invariants in $\Delta_*(\wedge^+\hat{P})$. Moreover a linear basis of $\wedge\hat{P}$ leads to linearly independent cohomology classes in $H(\underline{\underline{g}},H)$.

In case of a surjective map $I(G) \longrightarrow I(H)$ in fact $H(\underline{\underline{g}},H) \cong \wedge\hat{P}$ so that these are then the only invariants. This is the situation in several examples discussed below. See theorem 6.28 for trivial complex bundles with a real structure, theorem 6.33 for flat $GL(m)$-bundles $(H = O(m))$, theorem 6.49 for $SO(2m-1)$-bundles with trivial $SO(2m)$-extension and theorem 6.52 for trivial complex bundles with a symplectic structure.

We consider again flat bundles of the type

$$(6.10) \qquad P = \Gamma\backslash G \times_H G \cong G/H \times_\Gamma G \longrightarrow M = \Gamma\backslash G/H$$

for G a Lie group, $H \subset G$ a closed subgroup and $\Gamma \subset G$ a discrete subgroup operating properly discontinuously and without fixed points on G/H. The double coset space M is then a manifold. For compact G, the group Γ is finite. The characteristic homomorphism of chapter 3

$$(6.11) \qquad \Delta_*(P) = \gamma_* : H(\underline{\underline{g}},H) \longrightarrow H^{\cdot}_{DR}(M)$$

is induced by the canonical inclusion

$$(6.12) \qquad \gamma : (\wedge\underline{\underline{g}}^*)_H \longrightarrow \Omega^{\cdot}(\Gamma\backslash G/H).$$

6.13 THEOREM. _Let_ G _be a compact connected Lie group,_ $H \subset G$ _a closed subgroup and_ Γ _a finite subgroup of_ G _acting without fixed points on_ G/H. _The generalized characteristic homomorphism_

$$\Delta_* = \gamma_* : H(\underline{\underline{g}},H) \longrightarrow H_{DR}(M)$$

<u>of the flat bundle</u>

$$P = \Gamma\backslash G \times_H G \cong G/H \times_\Gamma G \to M = \Gamma\backslash G/H$$

<u>is an isomorphism.</u>

Proof. The realization of $H_{DR}(G/H)$ by G-invariant cohomology classes establishes the isomorphism $H(\underline{\underline{g}},H) \cong H_{DR}(G/H)$. The result follows, since the finite group $\Gamma \subset G$ acts trivially on $H_{DR}(G/H)$. □

For $\Gamma = \{e\}$ the map $\gamma : (\wedge^{\cdot}\underline{\underline{g}}^*)_H \longrightarrow \Omega^{\cdot}(G/H)$ is the inclusion of left-invariant forms into $\Omega^{\cdot}(G)$, restricted to H-basic elements. Thus e.g. for a compact connected group G the Chevalley-Eilenberg cohomology isomorphism $H(\underline{\underline{g}}) \xrightarrow{\cong} H(G)$ induced by the canonical inclusion of left-invariant forms $\wedge^{\cdot}\underline{\underline{g}}^* \longrightarrow \Omega^{\cdot}(G)$ can be interpreted as the generalized characteristic homomorphism of the trivial bundle $G \times G \longrightarrow G$. The foliation is given by the diagonal action of G on $G \times G$, which is non-compatible with the trivialization, and therefore gives rise to secondary invariants by the principle embodied in theorem 4.52.

Assume in particular G/H to be a symmetric space. It is then easy to check that the differential on $(\wedge\underline{\underline{g}}^*)_H$ is zero, so that in fact $\gamma_* : (\wedge\underline{\underline{g}}^*)_H \cong H(G/H)$. This can be applied to the symmetric space defined by the $G \times G$-action on G via $(g,g')\cdot g'' = gg''g'^{-1}$. In this case the diagonal ΔG plays the role of the subgroup H and by the remark above

$$\gamma_* : \wedge(\underline{\underline{g}} \times \underline{\underline{g}})^*_{\Delta\underline{\underline{g}}} \cong H(G)$$

The left-hand side represents the biinvariant forms on G with trivial differential and hence is isomorphic to H(G). This is

the interpretation of the generalized characteristic homomorphism
in this case.

We turn now to applications of Theorem 6.13 (G and H
compact). For $\Gamma = \{e\}$ note that P is the G-extension of the
H-bundle $G \longrightarrow G/H$. Let more generally $P' \longrightarrow M$ be an H-bundle
with trivial G-extension $P = P' \times_H G$. The fibration
$G/H \longrightarrow B_H \longrightarrow B_G$ shows that P' together with a trivialization
of P is characterized by a homotopy class $f : M \longrightarrow G/H$. The
diagram corresponding to (4.86) is in this case induced by the
commutative diagram of space maps

(6.14)

$$
\begin{array}{ccc}
B_G & \longleftarrow & B_H \\
{\scriptstyle g\simeq 0}\uparrow & \nearrow{\scriptstyle g'} & \uparrow \\
M & \xrightarrow{\quad f \quad} & G/H
\end{array}
$$

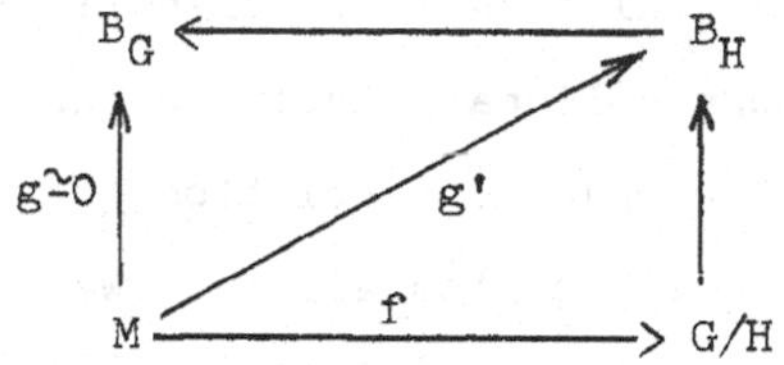

where g classifies P, g' classifies P'. The generalized
characteristic homomorphism $\Delta_*(P)$ of the trivial G-bundle P with
the H-reduction P' factorizes then as follows

(6.15)
$$
\Delta_* : H(\underline{g},H) \xrightarrow[\cong]{\gamma_*} H_{DR}(G/H) \xrightarrow{f_*} H_{DR}(M)
$$

and Δ_* depends only on the homotopy class of $f : M \longrightarrow G/H$.

A first application of this construction is to the case
$G = U(m)$ and $H = O(m)$. We need $I(U(m))$ and $I(O(m))$.

First recall that the (complex) invariant polynomials are
given by
$$
I(GL(m,\mathbb{C})) \cong \mathbb{C}[c_1,\ldots,c_m].
$$

The Chern polynomials $c_j \in I^{2j}(GL(m,\mathbb{C}))$ are defined by the identity

(6.16)
$$
\det(Id + \frac{t}{2\pi} A) = \sum_{j=0}^{m} c_j(A)t^j,
$$

where $c_j(A) = (\frac{1}{2\pi})^j$ trace $\wedge^j A$ for $A \in \underline{\underline{gl}}(m,\mathbb{C})$. For a triangular matrix A with diagonal elements $\lambda_1,\ldots,\lambda_m$ the identity

$$\det(\mathrm{Id} + \frac{t}{2\pi} A) = \prod_{j=1}^{m} (1 + \frac{t}{2\pi} \lambda_j) = \sum_{j=0}^{m} (\frac{1}{2\pi})^j \sigma_j (\lambda_1,\ldots,\lambda_m) t^j$$

shows that

$$c_j(A) = (\frac{1}{2\pi})^j \sigma_j (\lambda_1,\ldots,\lambda_m). \tag{6.17}$$

Here $\sigma_j(\lambda_1,\ldots,\lambda_m)$ denotes the j-th elementary symmetric function in $\lambda_1,\ldots,\lambda_m$.

To describe the real invariant polynomials $I(U(m))$, let $\tau(A) = -\bar{A}^t$ be the conjugate linear involution of $\underline{\underline{gl}}(m,\mathbb{C})$ leaving $\underline{\underline{u}}(m)$ fixed. The induced involution $\bar{\tau}$ of $I(GL(m,\mathbb{C}))$ is characterized by $\bar{\tau} c_j = (-1)^j c_j$ for all j. With the definition $\mathfrak{c}_j = i^j c_j$ we have then for the restriction to $\underline{\underline{u}}(m)$

$$I(U(m)) \cong \mathbb{R}[\mathfrak{c}_1,\ldots,\mathfrak{c}_m]. \tag{6.18}$$

The (modified) Chern polynomials $\mathfrak{c}_j \in I^{2j}(U(m))$ are characterized by the identity

$$\det(\mathrm{id} - \frac{t}{2\pi i} A) = \sum_{j=0}^{m} \mathfrak{c}_j(A)\, t^j \tag{6.19}$$

for $A \in \underline{\underline{u}}(m)$. With the purely imaginary eigenvalues $\lambda_j = i\mu_j$ of $A \in \underline{\underline{u}}(m)$ we get then from the identity

$$\det(\mathrm{Id} - \frac{t}{2\pi i} A) = \prod_{j=1}^{m} (1 - \frac{t}{2\pi} \mu_j) = \sum_{j=0}^{m} (-\frac{1}{2\pi})^j \sigma_j(\mu_1,\ldots,\mu_m) t^j$$

the expression

$$\mathfrak{c}_j(A) = (-\frac{1}{2\pi})^j \sigma_j(\mu_1, \ldots ,\mu_m). \tag{6.20}$$

The real form $\mathfrak{gl}(m) = \mathfrak{gl}(m,\mathbb{R})$ of $\mathfrak{gl}(m,\mathbb{C})$ is given as the fixed point set of the conjugate linear involution $\sigma_0(A) = \overline{A}$ of $\mathfrak{gl}(m,\mathbb{C})$. The induced involution $\overline{\sigma}_0$ of $I(GL(m,\mathbb{C}))$ leaves the c_j fixed, so that the real invariant polynomials on $GL(m) = GL(m,\mathbb{R})$ are given by

$$(6.21) \qquad I(GL(m)) \cong \mathbb{R}[c_1,\ldots,c_m]$$

The subalgebra $\mathfrak{so}(m) = \mathfrak{gl}(m) \cap \mathfrak{u}(m)$ is the set of common fixed points of the involutions σ_0 and τ. For the induced involutions we have then $\overline{\tau}_0 c_j = c_j$ in $I(GL(m))$ and $\overline{\tau}\mathfrak{c}_j = \mathfrak{c}_j$ in $I(U(m))$. From $\mathfrak{c}_j = i^j c_j$ it follows that under the restriction homomorphisms

$$I(U(m)) \longrightarrow I(O(m)) \longleftarrow I(GL(m))$$

the Chern polynomials $\mathfrak{c}_{2j-1} = i^{2j-1} c_{2j-1}$ resp. c_{2j-1} vanish for $0 < 2j-1 \leq m$. It follows that

$$(6.22) \qquad I(O(m)) \cong \mathbb{R}[p_1,\ldots,p_{[m/2]}]$$

where the Pontrjagin polynomials $p_j \in I^{4j}(O(m))$ are the restrictions of $(-1)^j\mathfrak{c}_{2j}$ to $\mathfrak{so}(m)$ for $j = 1,\ldots,[\frac{m}{2}]$. The identity (6.19) restricts to

$$(6.23) \qquad \det(Id - \frac{t}{2\pi i} A) = \sum_{j=0}^{[m/2]} p_j(A)\, t^{2j}\, (-1)^j$$

on $\mathfrak{so}(m)$. The eigenvalues of the complexification of $A \in \mathfrak{so}(m)$ are $i\mu_1,\ldots,i\mu_{[m/2]}, -i\mu_1,\ldots, i\mu_{[m/2]}$ with real μ_j's, plus a zero if m is odd. Thus

$$\det(\mathrm{Id} - \frac{t}{2\pi i} A) = \prod_{j=1}^{[m/2]} (1+\mu_j \frac{t}{2\pi})(1-\mu_j \frac{t}{2\pi}) = \prod_{j=1}^{[m/2]} (1-\mu_j^2(\frac{t}{2\pi})^2)$$

$$= \sum_{j=0}^{[m/2]} \frac{(-1)^j}{(2\pi)^{2j}} \sigma_j(\mu_1^2, \ldots, \mu_m^2) t^{2j}$$

so that

(6.24) $$p_j(A) = \frac{1}{(2\pi)^{2j}} \cdot \sigma_j(\mu_1^2, \ldots, \mu_{[m/2]}^2).$$

After these generalities, we turn to the determination of $H(\underline{\underline{u}}(m), O(m))$ using the isomorphism (6.6).

The restriction $i^* : I(U(m)) \longrightarrow I(O(m))$ is characterized by

(6.25) $$i^*\tilde{c}_{2j-1} = 0 \quad \text{and} \quad i^*\tilde{c}_{2j} = (-1)^j p_j \quad (j = 1, \ldots, [\tfrac{m}{2}]).$$

Let $y_j = \sigma\tilde{c}_j$ be the primitive generators for $\underline{\underline{u}}(m)$. Then the Samelson space $\hat{P} \subset (\Lambda \underline{\underline{u}}(m)^*)_{O(m)}$ is by (6.5) spanned by $y_1, y_3, \ldots, y_{m'}$, where $m' = 2[\frac{m+1}{2}] - 1$ is the largest odd integer $\leq m$. It follows from (6.6) that

(6.26) $$H(\underline{\underline{u}}(m), O(m)) \cong \Lambda\hat{P} \cong \Lambda(y_1, y_3, \ldots, y_{m'})$$

By (6.20) and the formula (5.74) for the suspension of σc_1 we have e.g. the explicit formula

(6.27) $$y_1 = \frac{i}{2\pi} \operatorname{trace} \in (\Lambda^1 \underline{\underline{u}}(m)^*)_{O(m)}.$$

6.28 THEOREM [KT 9,10]. <u>Let</u> $P' \longrightarrow M$ <u>be an</u> $O(m)$-<u>bundle with a trivial</u> $U(m)$-<u>extension</u> (<u>a trivial complex bundle with a real structure</u>). <u>Then there are well-defined secondary characteristic invariants</u>

$\Delta_*(y_i) \in H_{DR}^{2i-1}(M)$ for $i = 1, 3, \ldots, m'$, $m' = 2[\frac{m+1}{2}] - 1$.

These invariants are according to Theorem 4.52 obstructions to the triviality of the real structure on the trivial complex vectorbundle.

To give an interpretation of $\Delta_*(y_1)$, we refer now to the cohomology class introduced by Maslov [MS] and which intervenes in quantization conditions (see the discussion by Arnold in [A], the same in an appendix to [MS]).

6.29 PROPOSITION [KT 9,10]. Let $P' \longrightarrow M$ be an $O(m)$-bundle with a trivial $U(m)$-extension P, ω the connection form of the trivial connection in P. Then $\Delta_*(y_1)$ is represented by a closed 1-form $\Delta(\omega)(y_1)$ on M. The Maslov class of P' is the characteristic class $-2\Delta_*(y_1) \in H^1_{DR}(M)$ and

$$-2\int_\gamma \Delta(\omega)(y_1) = \deg\,(\det{}^2 \circ f(\gamma)) \quad \text{for} \quad \gamma \in \pi_1(M).$$

Proof. The map $\det : U(m) \longrightarrow S^1$ squared factorizes through $O(m)$ and defines $\det^2 : U(m)/O(m) \longrightarrow S^1$. The RHS is then the degree of the map $S^1 \xrightarrow{f(\gamma)} U(m)/O(m) \xrightarrow{\det^2} S^1$, where $f : M \longrightarrow U(m)/O(m)$ classifies P' with its trivialized $U(m)$-extension.

This is proved by observing that for the $O(m)$-reduction $\bar{s} : P' \longrightarrow P$ given by $s : M \longrightarrow P/O(m)$, we can represent $\Delta(\omega)(y_1)$ as the $O(m)$-basic 1-form $\frac{i}{2\pi}\,\bar{s}^*\,(\text{trace}\ \omega)$ on P'. It suffices to check the formula for the critical example $O(m) \to U(m) \to U(m)/O(m) = M$. Observe further that for a lift $\tilde{\gamma}$ to $U(m)$ of $\gamma \in \pi_1(U(m)/O(m))$ clearly $\int_{\tilde{\gamma}} \bar{s}^*\ \text{trace}\ \omega = \int_\gamma s^*\ \text{trace}\ \omega$. Since $\pi_1(U(m)/O(m)) \cong \mathbb{Z}$, it suffices to verify that

$$-\frac{i}{\pi}\int_\gamma \bar{s}^*\ \text{trace}\ \omega = \deg\,(\det{}^2 \circ \gamma)$$

for a single path $\gamma : [0,2\pi] \longrightarrow U(m)$ which maps into a nontrivial loop in $\pi_1(U(m)/O(m))$. For the path $\gamma(t) = e^{it/2}$ in $U(m)$ it is then easily verified that the value of both terms is m, which completes the proof. □

At this place we wish to discuss the non-compact version of these classes. Let P be a flat $GL(m)$-bundle. It has an $O(m)$-reduction and the generalized characteristic homomorphism does not depend on the choice of the $O(m)$-reduction, since there is only one homotopy class of sections of $P/O(m) \longrightarrow M$. By (6.21) we have $I(GL(m)) \cong \mathbb{R}[c_1,\ldots,c_m]$, where the c_j are the restrictions of the Chern polynomials to $\underline{\underline{gl}}(m) = \underline{\underline{gl}}(m,\mathbb{R})$. The restriction $i^* : I(GL(m)) \longrightarrow I(O(m))$ is characterized by

$$(6.30) \qquad i^*c_{2j-1} = 0, \quad i^*c_{2j} = p_j \qquad (j=1,\ldots,[\tfrac{m}{2}])$$

With $y_j = \sigma c_j$ the Samelson space $\hat{P} \subset (\wedge\, \underline{\underline{gl}}(m)^*)_{O(m)}$ is the space spanned by $y_1, y_3, \ldots, y_{m'}$ (m' the largest odd integer $\leq m$) and by (6.6) therefore

$$(6.31) \qquad H(\underline{\underline{gl}}(m), O(m)) \cong \wedge(y_1, y_3, \ldots, y_{m'}).$$

Note that with these normalizations e.g.

$$(6.32) \qquad y_1 = \frac{1}{2\pi}\, \mathrm{trace} \in (\wedge^1 \underline{\underline{gl}}(m)^*)_{O(m)}.$$

6.33 THEOREM [KT 7, thm. 4.5]. Let $P \longrightarrow M$ be a flat $GL(m)$-bundle. There are well-defined secondary characteristic invariants $\Delta_*(y_i) \in H^{2i-1}_{DR}(M)$ for $i = 1,3,\ldots,m'$, $m' = 2[\frac{m+1}{2}] - 1$. If P is a flat $O(m)$-bundle, these invariants are zero.

By theorem 4.52 the non-triviality of the classes $\Delta_*(y_i)$ is a measure for the incompatibility of the flat $GL(m)$-structure with the (up to homotopy) canonical $O(m)$-reduction.

These classes are closely related to the invariants defined by Reinhart [RE 3] and Goldman [GL 1,2] on a leaf of a foliation. The foliated normal bundle restricted to a leaf is flat, so that it carries the invariants described above.

Since the flat bundle P is completely characterized by the holonomy representation $h : \pi_1(M) \longrightarrow GL(m)$, it is interesting to determine the invariants $\Delta_*(y_i)$ from h. For the invariants $\Delta_*(y_1) \in H^1_{DR}(M)$ this is done by the following formula [KT 7, Theorem 4.5].

6.34 PROPOSITION. Let P be a flat GL(m)-bundle with connection form ω. Then $\Delta_*(y_1)$ is represented by a closed 1-form $\Delta(\omega)(y_1)$ on M and

$$\int_\gamma \Delta(\omega)(y_1) = -\frac{1}{2\pi} \log |\det h(\gamma))| \text{ for } \gamma \in \pi_1(M).$$

This formula shows that $\Delta_*(y_1)$ is non-zero if and only if the holomony representation does not map into the $(m \times m)$-matrices with determinant ± 1.

Note that by the formula for $\Delta_*(y_1)$ this invariant is visibly not invariant under deformations, whereas this is the case for all invariants $\Delta_*(y_i)$ for $i > 1$. The subject of the deformation invariance of generalized characteristic classes has only been touched in these lectures (see 4.75). The appropriate framework has been sketched in [KT 7]. For the present context the relevant rigidity result is theorem 8.11 of [KT 7]. It implies the rigidity of the classes $\Delta_*(y_i)$ for $i > 1$. For the simplest case of a projection $M = X \times F \longrightarrow X$ the relevant statement is the commutative diagram (4.76) with $q = 0$.

Proof of proposition 6.34. Let $s : M \longrightarrow P/O(m)$ define the $O(m)$-reduction P' of P. Then by (6.32)

$$\Delta(\omega)(y_1) = \frac{1}{2\pi} s^* \operatorname{trace} \omega \in \Omega^1(M)$$

so that we have to show

$$(6.35) \qquad \int_\gamma s^* \operatorname{trace} \omega = -\log|\det h(\gamma)| \text{ for } \gamma \in \pi_1(M).$$

Consider the homomorphism $|\det| : GL(m) \longrightarrow \mathbb{R}^{*+}$ and the corresponding bundle map

$$P \longrightarrow \bar{P} = P \times_{GL(m)} \mathbb{R}^{*+}$$

The flat connection with holonomy h defined by the connection form ω in P gives rise to a flat connection with holonomy $\bar{h} = |\det h|$ defined by the connection form $\bar{\omega}$ on $\bar{P}$. The homomorphism $O(m) \longrightarrow GL(m) \xrightarrow{|\det|} \mathbb{R}^{*+}$ is trivial. Thus the section s of $P/O(m) \longrightarrow M$ defines a section $\bar{s}$ of $\bar{P} \longrightarrow M$. By the functoriality of Δ_* as discussed in 4.59, it follows that for every linear function $\alpha : \mathbb{R} \longrightarrow \mathbb{R}$

$$\bar{s}^*(\alpha\bar{\omega}) = s^*(W(D(|\det|))\alpha)(\omega).$$

But $D(|\det|) = \operatorname{trace} : \underline{\underline{gl}}(m) \longrightarrow \mathbb{R}$ and $W(D(|\det|))\alpha = \operatorname{trace}^* \alpha = \alpha \circ \operatorname{trace}$. It follows that

$$\bar{s}^*(\alpha\bar{\omega}) = s^*(\alpha \operatorname{trace} \omega)$$

and in particular for $\alpha = \mathrm{id} : \mathbb{R} \longrightarrow \mathbb{R}$

$$\bar{s}^*\bar{\omega} = s^*(\operatorname{trace} \omega).$$

The identity (6.35) translates therefore to

$$\int_\gamma \bar{s}^*\bar{\omega} = -\log \bar{h} \quad \text{for} \quad \gamma \in \pi_1(M).$$

Note that for $\bar{s}$ we can take any section of the $\mathbb{R}^{*+}$-bundle $\bar{P}$. Simplifying notations, the problem reduces to the verification of the formula

$$(6.36) \qquad \int_\gamma s^*\omega = -\log h(\gamma) \quad \text{for} \quad \gamma \in \pi_1(M)$$

for any flat connection ω with holonomy $h : \pi_1(M) \longrightarrow \mathbb{R}^{*+}$ in a $\mathbb{R}^{*+}$-bundle $P \longrightarrow M$ with section s.

Choose a basepoint $x_0 \in M$ and let $p_0 = s(x_0)$. For $\gamma \in \pi_1(M,x_0)$ let $\tilde{\gamma}$ be the lift in P with respect to the flat connection ω with initial point p_0. Then $\tilde{\gamma}(1) = \tilde{\gamma}(0) \cdot h(\gamma)$ by the definition of holonomy. If γ is parametrized by $[0,1]$, then more generally

$$(6.37) \qquad \tilde{\gamma}(t) = s(\gamma(t)) \cdot \lambda(t)$$

where $\lambda : [0,1] \longrightarrow \mathbb{R}^{*+}$. In the canonically associated line bundle $L \longrightarrow M$ this formula reads $\tilde{\gamma} = \lambda s$ for the corresponding sections of L. By construction $\nabla_{\dot{\gamma}}\tilde{\gamma} = 0$, so that

$$(6.38) \qquad \nabla_{\dot{\gamma}}(\lambda s) = \lambda \nabla_{\dot{\gamma}} s + \dot{\gamma}\lambda \cdot s = 0$$

But with the above identification of s as a section of P and of L we have clearly $\nabla_{\dot{\gamma}} s = (s^*\omega)(\dot{\gamma}) \cdot s$, so that (6.38) implies

$$\lambda . (s^*\omega)(\dot{\gamma}) + \dot{\lambda} = 0$$

The solution is given by

$$\log \lambda(t) = -\int_0^t (s^*\omega)(\dot{\gamma}(\tau))d\tau + c$$

But by (6.37) we have $\lambda(0) = 1$, hence $\log \lambda(0) = 0$ and $c = 0$. Further $\lambda(1) = h(\gamma)$, so that

$$\log h(\gamma) = -\int_0^1 (s^*\omega)(\dot\gamma(\tau))d\tau = -\int_\gamma s^*\omega$$

which is the desired formula (6.36). □

In the following situation we obtain a non-trivial realization of the invariant $\Delta_*(y_1)$.

6.39 PROPOSITION [KT 10]. Let M^m be a compact affine hyperbolic manifold. Then for the tangentbundle $\Delta_*(y_1)$ is a non-trivial cohomology class.

Proof. The hyperbolicity of the affine structure means that the universal covering of M^m is affinely isomorphic to an open convex subset of R^m containing no complete line (these are non-complete affine manifolds). According to Koszul [K 5], the hyperbolicity condition is equivalent to the existence of a closed 1-form β with positive definite covariant derivative $\nabla\beta$. Note that $g(X,Y) = (\nabla_X\beta)(Y) = X\beta(Y) - \beta(\nabla_X Y)$ is a symmetric form. This follows from

$$g(X,Y) - g(Y,X) = X\beta(Y) - Y\beta(X) - \beta(\nabla_X Y - \nabla_Y X) = d\beta(X \wedge Y) - \beta(T(X,Y)) = 0$$

since the torsion T of the flat connection is zero. Koszul's definition may be used to show that the 1-form $\Delta(\omega)(y_1)$ can be identified with β.

Assume now $\Delta_*(y_1)$ to be zero in $H^1(M)$. Then $\Delta(\omega)(y_1) = df$ for a smooth function $f: M \longrightarrow \mathbb{R}$. Therefore $g(X,Y) = (\nabla_X df)(Y)$. Let $x \in M$ be a critical point of f. For any curve γ with initial point $\gamma(0) = x$ we have

$$g(\dot\gamma,\dot\gamma) = (\nabla_{\dot\gamma} df)(\dot\gamma) = \dot\gamma\, df(\dot\gamma) - df(\nabla_{\dot\gamma}\dot\gamma).$$

For $t = 0$ we get therefore

$$g(\dot{\gamma}(0),\dot{\gamma}(0)) = \frac{d^2f}{dt^2}(\gamma(t))\Big|_{t=0} .$$

This shows that g is the Hessian of f in every initial point. But M is compact and the Hessian of f in a point where f attains its maximum is not positive definite. This contradiction shows that $\Delta_*(y_1)$ is a non-trivial cohomology class. □

We turn to a further application of the preceding constructions. Let $G = SO(2m)$ and $H = SO(2m-1)$, so that $G/H = S^{2m-1}$. For an $SO(2m-1)$-bundle $P' \longrightarrow M$ with trivial $SO(2m)$-extension P there is a unique homotopy class $f : M \longrightarrow S^{2m-1}$. Then $\ker(I(SO(2m)) \longrightarrow I(SO(2m-1))$ is generated by the (normalized) Pfaffian polynomial $e_m \in I^{2m}(SO(2m))$ which is defined as follows. Let $x_1,\dots,x_{2m}$ be an orthonormal basis of $\mathbb{R}^{2m}$ with dual basis $x_1^*,\dots,x_{2m}^*$ and define

(6.40) $$\omega(A) = \sum_{i<j} A_{ij}\, x_i^* \wedge x_j^* \in \wedge^2 \mathbb{R}^{2m*}$$

for $A = (A_{ij}) \in \underline{\underline{so}}(2m)$. The 2-form $\omega(A)$ is $SO(2m)$-invariant and hence $\omega^m(A)$ is an $SO(2m)$-invariant volume element on $\mathbb{R}^{2m}$. As such it is a multiple of $\mu = x_1^* \wedge \dots \wedge x_{2m}^*$, and therefore $\omega^m(A) = (-1)^m . (2\pi)^m . m! . e_m(A) . \mu$ defines $e_m \in I^{2m}(SO(2m))$. The explicit formula for e_m in terms of a basis of $S^1(\underline{\underline{so}}(2m)^*)$ is

(6.41) $$e_m = \frac{(-1)^m}{(2\pi)^m m!} \cdot \sum_{\substack{i_k<j_k \\ k=1,\dots,m}} \varepsilon_\sigma\, \tilde{X}^*_{i_1j_1} \cdots \tilde{X}^*_{i_mj_m}$$

where $\varepsilon_\sigma = \operatorname{sign}\sigma$, $\sigma(1\dots 2m) = (i_1j_1 \dots i_mj_m)$. $X_{ij}(i<j)$ is the basis element of $\underline{\underline{so}}(2m)$ with a 1 in the i-th row and j-th column, -1 in the j-th row and i-th column and zero

everywhere else. X^*_{ij} denotes the dual basis of $\underline{\underline{so}}(2m)^*$.

With this notation

(6.42) $$I(SO(2m)) \cong \mathbb{R}[p_1,\ldots,p_{m-1},e]; \quad p_m = e^2,$$

(6.43) $$I(SO(2m-1) \cong \mathbb{R}[p_1,\ldots,p_{m-1}],$$

and the restriction map $i^* : I(SO(2m)) \longrightarrow I(SO(2m-1))$ is given by

(6.44) $$i^*p_j = p_j \quad \text{for} \quad j = 1,\ldots,m-1 \quad \text{and} \quad i^*e = 0.$$

Therefore $\hat{P}$ is spanned by

(6.45) $$\sigma e \in (\Lambda^{2m-1}\, \underline{\underline{so}}(2m)^*)_{SO(2m-1)}$$

and by (6.6)

(6.46) $$H(\underline{\underline{so}}(2m), SO(2m-1)) \cong \Lambda(\sigma e) \cong \mathbb{R}[\sigma e]/(\sigma e)^2.$$

This is of course the cohomology of $SO(2m)/SO(2m-1) \cong S^{2m-1}$. The generator σe in top degree is evaluated as follows.

Let $\underline{\underline{so}}(2m) = \underline{\underline{so}}(2m-1) \oplus \underline{\underline{m}}$ be the symmetric space decomposition with $\underline{\underline{m}}$ spanned by the matrices $Z_\alpha = X_{\alpha,2m}$ $(\alpha = 1,\ldots,2m-1)$. Then

$$\sigma(e) \in (\Lambda^{2m-1}\underline{\underline{so}}(2m)^*)_{SO(2m-1)} \cong (\Lambda^{2m-1}\underline{\underline{m}}^*)^{SO(2m-1)}$$

is a multiple of the volume form

$$\mu = Z^*_1 \wedge \ldots \wedge Z^*_{2m-1} \in (\Lambda^{2m-1}\underline{\underline{m}}^*)^{SO(2m-1)}.$$

6.47 PROPOSITION [KT 9,10]. <u>Let the notation be as above. Then</u> $\sigma(e) = \alpha.\,\mu$ <u>with</u>

(6.48) $$\alpha = -\frac{(m-1)!}{(2\pi)^m 2^{m-1}(2m-1)}.$$

Proof. This follows by a direct computation using the fact that the volume of S^{2m-1} with respect to the invariant form μ is $\frac{2}{(m-1)!}\cdot\pi^m$. □

To give a geometric interpretation to the invariant $\sigma(e)$, consider the following situation.

6.49 THEOREM [KT 9,10]. Let $h : M^{2m-1} \longrightarrow \mathbb{R}^{2m}$ be an isometric immersion of the compact oriented Riemannian manifold M. The tangent frame bundle $SO(2m-1) \longrightarrow P' \longrightarrow M$ has a trivial $SO(2m)$-extension $P \longrightarrow M$. The generalized characteristic homomorphism $\Delta_* = \Delta_*(P)$ applied to $\sigma(e)$ gives a top-dimensional cohomology class on M such that

$$N(h) = -2^{2(m-1)}(2m-1)\Delta_*\,\sigma(e)[M]$$

where $N(h)$ is the normal degree of h.

Proof. The $SO(2m-1)$-bundle P' with trivial $SO(2m)$-extension P is characterized by a map $f : M^{2m-1} \longrightarrow S^{2m-1}$ which (up to homotopy) is precisely the Gauss map g_h of h. Thus with the previous notations $\Delta_* = g_h^* \circ \gamma_*$ and in top degree $2m-1$ in particular $\Delta_* = \deg(g_h)\cdot\gamma_* \equiv N(h).\gamma_*$. This establishes the functoriality of $\Delta_*\sigma(e)$ in the sense that

$$\Delta_*\sigma(e)[M^{2m-1}] = N(h).\Delta_*\sigma(e)[S^{2m-1}],$$

and it suffices to verify that

$$\Delta_*\sigma(e)[S^{2m-1}] = \int_{S^{2m-1}} \gamma_*\sigma(e) = -\frac{1}{2^{2(m-1)}(2m-1)} .$$

But with α as in 6.47

$$\int_{S^{2m-1}} \gamma_*\sigma(e) = \alpha \int_{S^{2m-1}} \mu$$

and $\int_{S^{2m-1}} \mu = \frac{2}{(m-1)!} \cdot \pi^m$, which proves the desired result. □

For an immersion h of S^1 in $\mathbb{R}^2$ the formula above shows that the normal degree

$$(6.50) \qquad N(h) = -\Delta_* \sigma(e)[S^1]$$

where $\Delta_* \sigma(e)$ is computed with respect to the pullback metric on S^1 by h. This is of course the rotation index of the immersion h (up to sign).

It is of interest to contrast the last theorem with the situation when an even-dimensional compact manifold M^{2m} is immersed in $\mathbb{R}^{2m+1}$. For such an immersion $h : M^{2m} \longrightarrow \mathbb{R}^{2m+1}$ by Hopf [HO 1,2] the normal degree is given by $N(h) = \frac{1}{2}\chi(M)$, where $\chi(M)$ is the Euler number of M. In the framework of our discussion this formula follows from: (a) the functoriality

$$\Delta_* e[M] = N(h).\Delta_* e[S^{2m}]$$

of the primary invariant $\Delta_* e$ (e the Pfaffian polynomial in $I^{2m}(SO(2m))$, and (b) the evaluation

$$\Delta_* e[S^{2m}] = \int_{S^{2m}} \gamma_* e = 2$$

of this invariant on the sphere S^{2m}. Note that in this case the restriction map $i^* : I(SO(2m+1)) \longrightarrow I(SO(2m))$ is given by

$$(6.51) \qquad i^* p_j = p_j \quad \text{for} \quad j = 1,\dots,m$$

where $p_m = e^2$ in $I(SO(2m))$. The Samelson space is zero and (6.6) implies

$$H(\underline{\underline{so}}(2m+1), SO(2m)) \cong \mathbb{R}[e]/(e^2)$$

which is of course the cohomology of $SO(2m+1)/SO(2m) \cong S^{2m}$ generated by the Euler class.

In contrast to the primary nature of the Euler class (its definition is independent of the immersion), our secondary invariant $\Delta_* \sigma(e)$ needs for its definition on M^{2m-1}, besides the Riemannian structure, a trivialization of $\tau_M \oplus \varepsilon_1$, where τ_M is the tangent bundle of M and ε_1 a trivial line bundle on S^{2m-1}. It is in fact an invariant defined for Riemannian π-manifolds M^{2m-1} and allows to test the Riemannian immersability of M^{2m-1} in $\mathbb{R}^{2m}$. Note that Theorem 6.49 holds also for an isometric immersion $h : M^{2m-1} \longrightarrow N^{2m}$ into a Riemannian parallelizable manifold N^{2m} (the Gauss map g_h and the normal degree $N(h) = \deg(g_h)$ are then defined).

A final application is to Sp(m)-bundles with trivial U(2m)-extension. SP(m) denotes the compact group

$$Sp(m) = Sp(m,\mathbb{C}) \cap U(2m) \subset GL(2m,\mathbb{C})$$

Its Lie algebra $\underline{\underline{sp}}(m)$ consists of the matrices $A \in \underline{\underline{gl}}(2m,\mathbb{C})$ such that

$$JA = -A^t J \quad \text{and} \quad A = -\bar{A}^t$$

where J is the $(2n \times 2n)$-matrix $\left[\begin{array}{c|c} 0 & Id \\ \hline -Id & 0 \end{array}\right]$. The symplectic Pontrjagin polynomials $e_j \in I^{Mj}(Sp(m))$ are the restrictions of the $\tilde{c}_{2j} \in I^{4j}(U(2m))$ to $\underline{\underline{sp}}(m)$ and $\ker(I(U(2m)) \longrightarrow I(Sp(m))$ is spanned by the $\tilde{c}_i$ for $i = 1,3,\ldots,m' = 2[\frac{m+1}{2}] - 1$. With $\sigma\tilde{c}_j = y_j$ by (6.6) then

$$H(\underline{\underline{u}}(2m), Sp(m)) \cong \wedge(y_1, y_3, \ldots, y_{m'})$$

In complete analogy to (6.28) we have then the following result.

6.52 THEOREM. *Let* $P' \longrightarrow M$ *be an* $Sp(m)$*-bundle with a trivial* $U(2m)$*-extension (a trivial complex bundle with a symplectic structure. Then there are well-defined secondary invariants*

$$\Delta_*(y_i) \in H_{DR}^{2i-1}(M) \quad \text{for} \quad i = 1,3,\ldots,m' = 2[\tfrac{m+1}{2}] - 1.$$

These invariants are obstructions to the triviality of the symplectic structure.

7. EXAMPLES OF GENERALIZED CHARACTERISTIC CLASSES FOR FOLIATED BUNDLES

7.0 INTRODUCTION. In this chapter we discuss various examples of foliated bundles with non-trivial generalized characteristic invariants. After the computation of the Godbillon-Vey class we turn in particular to the characteristic classes of homomogeneous foliated bundles. The computation of Δ_* reduces then to purely algebraic problems which can be solved with the methods of chapter 5. According to these results, Δ_* for a locally homogeneous foliated vectorbundle is the composition of three maps. The first map is associated to the representation of (G,H) which defines a foliated vectorbundle in terms of a foliated principal bundle. The second map is expressible purely in terms of relative Lie algebra and Weil algebra cohomology. The third map is the characteristic homomorphism of a flat bundle, as discussed in Chapter 3. The precise statement are theorems 7.35 and 7.39. This evaluation principle is the basis of the results in [KT 9,10] and further papers to appear. The characteristic homomorphism of the foliation of a group by the cosets of a subgroup is computed explicitly in many cases. The techniques have to be varied slightly according to the assumptions on the pair $(\overline{G},G)$. Some of the results are given here, but we have to refer to [KT 10] for the technical details, which are too long to be reproduced in these notes.

7.1 DETERMINIATION OF $H(W(\underline{\underline{gl}}(m),O(m))_q)$. We need the restriction $i^* : I(GL(m)) \longrightarrow I(O(m))$, characterized in (6.30) by $i^*c_{2j-1} = 0$ and $i^*c_{2j} = p_j$ for $j = 1,\ldots,[\frac{m}{2}]$. The Samelson space $\hat{P} \subset (\Lambda\underline{\underline{gl}}(m)^*)_{O(m)}$ is spanned by the suspensions $y_1, y_3, \ldots, y_{m'}$, where $y_j = \sigma c_j$ and $m' = 2[\frac{m+1}{2}] - 1$ is the largest odd integer

$\leqq m$. Since $I(GL(m)) \longrightarrow I(O(m))$ is surjective, by theorem 5.108 $H(W(\underline{\underline{gl}}(m),O(m))_q)$ is the cohomology of the DG-algebra

$$(7.2) \qquad \wedge(y_1,y_3,\ldots,y_{m'}) \otimes \mathbb{R}[c_1,\ldots,c_m]_q$$

with differential characterized by $dy_j = c_j$ and $dc_j = 0$. By theorem 5.110 a basis for the secondary characteristic classes is given by the cocycles

$$(7.3) \qquad z_{(i,j)} = y_{i_1}\wedge\ldots\wedge y_{i_s} \otimes c_1^{j_1}\cdot\ldots\cdot c_m^{j_m}$$

with $1 \leq i_1 < \ldots < i_s \leqq m'$, i_ℓ odd, $s > 0$, $i_1 + p \geqq q+1$ for $p = \sum_{\ell=1}^{m} j_\ell \cdot \ell \leq q$ and $j_k = 0$ for $k < i_1, k$ odd. Note that

$$(7.4) \qquad 2q+1 \leqq \deg z_{(i,j)} \leqq 2q+m^2.$$

This basis has been computed for $m = q$ by Vey [GB 2]. The results of [KT 5] in chapter 5 determine $H(W(\underline{g},H)_q)$ for all indices q and all reductive pairs satisfying the special Cartan condition (CS) in (5.101). Among the cocycles of minimum degree $2q+1$ are those of the form

$$y_1 \otimes \Phi$$

where $\Phi(c_1,\ldots,c_m)$ is any polynomial of degree q as e.g. c_1^q. The class defined by

$$(7.5) \qquad y_1 \otimes c_1^q$$

is the Godbillon-Vey class [GV]. The cocycles of maximum degree $2q+m^2$ are given by

$$y_1 \wedge y_3 \wedge \ldots \wedge y_{q'} \otimes \Phi$$

where $\Phi(c_1,\ldots,c_m)$ is any polynomial of degree q. The classes

$z_{(i,j)}$ such that $dz_{(i,j)} \in F^{2k}$ for $k > q+1$ are invariant under deformations of the foliated bundle, whereas the classes $z_{(i,j)}$ satisfying $dz_{(i,j)} \in F^{2(q+1)}$ but $dz_{(i,j)} \notin F^{2k}$ for $k > q+1$ can be deformed in a non-trivial way (cf. 4.75 and also [HT]).

7.7 GODBILLON-VEY CLASS. For a codimension q foliation $L \subset T_M$ there is by theorem 4.43 a characteristic homomorphism

$$\Delta(F(Q^*))_*: H^{\cdot}(W(\underline{\underline{gl}}(q), O(q))_q) \longrightarrow H^{\cdot}_{DR}(M)$$

where $F(Q^*)$ is the foliated frame bundle of the dual Q^* of the normal bundle $Q = T_M/L$. The particular cohomology class

$$(7.8) \qquad \Delta(F(Q^*))_*(y_1 \otimes c_1^q) \in H^{2q+1}_{DR}(M)$$

is the Godbillon-Vey class of the foliation [GV]. Its non-triviality obstructs by theorem 4.52 the foliation in $F(Q^*)$ to be induced from a foliation in the orthogonal frame bundle of Q^*. We describe this class in more detail.

The homomorphism $\det : GL(q) \longrightarrow \mathbb{R}^*$ induces from $F(Q^*)$ the bundle

$$(7.9) \qquad \det_* F(Q^*) = F(\wedge^q Q^*)$$

It is the frame bundle of the foliated line bundle $\wedge^q Q^*$. Since the Godbillon-Vey class is already a class

$$(7.10) \qquad [y_1 \otimes c_1^q] \in H^{2q+1}(W(\underline{\underline{gl}}(1), O(1))_q)$$

it follows from the functoriality of the characteristic homomorphism Δ_* (see 4.43 and 4.59), that the Godbillon-Vey class (7.8) is also given by

$$(7.11) \qquad \Delta(F(\wedge^q Q^*))_*(y_1 \otimes c_1^q)$$

In the following we assume that $\wedge^q Q^*$ is trivial (i.e. Q^* is orientable). With respect to a trivialization s of $F(\wedge^q Q^*)$

the homomorphism $\Delta(F(\Lambda^q Q^*))_*$ is then realized on the cochain level by a map

$$(7.12) \qquad \Delta(\omega) : W(\underline{\underline{gl}}(1))_q \longrightarrow \Omega^{\cdot}(M).$$

and on the corresponding A-complex we have

$$(7.13) \qquad \tilde{\Delta}(\omega)(c_1) = h(\omega)(c_1) \in \Omega^2(M)$$

where $h : I(GL(1)) \longrightarrow \Omega(M)$ is the Chern-Weil homomorphism of the connection ω on $\Lambda^q Q^*$. By multiplicativity then $\tilde{\Delta}(\omega)(c_1^q) = (h(\omega)(c_1))^q \in \Omega^{2q}(M)$.

By (5.95), (iii) we have further

$$(7.15) \qquad \tilde{\Delta}(\omega)(y_1) = -\lambda^1(\omega,\omega')(c_1) \in \Omega^1(M)$$

where $\lambda^1(\omega,\omega') : I(GL(1)) \longrightarrow \Omega(M)$ is the difference map defined by (5.96). Since in the present case $H = \{e\}$ and $i^* = 0$ it follows by (5.97) that

$$(7.16) \qquad d\lambda^1(\omega,\omega')c_1 = -h(\omega)(c_1)$$

With the notation

$$(7.17) \qquad g = -\lambda^1(\omega,\omega')(c_1) \in \Omega^1(M)$$

one obtains then by multiplicativity for the Godbillon-Vey class the formula

$$(7.18) \qquad \tilde{\Delta}(\omega)(y_1 \otimes c_1^q) = g \wedge (dg)^q \in \Omega^{2q+1}(M)$$

Next we show that

$$(7.19) \qquad g = \frac{1}{2\pi} s^*\omega \in \Omega^1(M),$$

where s is a trivialization of $F(\Lambda^q(Q^*)$ pulling back the $\mathbb{R}$-valued adapted connection form ω to a 1-form on M. For this purpose observe again that for $H = \{e\}$ the map $\lambda^1(\omega,\omega')$ is by (5.96) the composition

$$I(\mathbb{R}) \xrightarrow{\lambda^1} W(\mathbb{R}) \otimes W(\mathbb{R}) \xrightarrow{(k(\omega),1)} \Omega(F(\wedge^q Q^*)) \xrightarrow{s^*} \Omega(M)$$

For $\tilde{\gamma} \in I^2(\mathbb{R}) \in S^1\mathbb{R}^*$ corresponding to $\gamma \in \wedge^1\mathbb{R}^*$ we have further by (5.55) the formula $\lambda^1\tilde{\gamma} = \varepsilon_0(\gamma) - \varepsilon_1(\gamma)$. From the particular form of $\lambda'(\omega,\omega')$ it follows that

$$\lambda^1(\omega,\omega')\tilde{\gamma} = -s^*\omega(\gamma)$$

For the polynomial $c_1 = \frac{1}{2\pi}\text{ trace} = \frac{1}{2\pi}\,.\,\text{id} : \mathbb{R}^* \longrightarrow \mathbb{R}^*$ we get then

$$\lambda^1(\omega,\omega')(c_1) = -\frac{1}{2\pi}\, s^*\omega$$

Together with (7.17) this proves (7.19).

If we consider the connection ∇ in $\wedge^q Q^*$ corresponding to ω in $F(\wedge^q Q^*)$, then the covariant derivative of the section ν of $\wedge^q Q^*$ corresponding to s of $F(\wedge^q Q^*)$ is given by

$$\nabla_X \nu = (s^*\omega)(X).\nu$$

for a tangent vectorfield X on M. It follows by (7.19) that g is characterized by the identity

$$\nabla_X \nu = 2\pi\, g(X).\nu$$

We summarize these calculations.

7.20 THEOREM (Godbillon-Vey). Let $L \subset T_M$ be a codimension q foliation with orientable normal bundle. Let ∇ denote an adapted connection in $\wedge^q Q^*$ and ν a section of $\wedge^q Q^*$. Then the Godbillon-Vey class $\Delta_*(Q^*)(y_1 \otimes c_1^q) \in H_{DR}^{2q+1}(M)$ is realized by the form $g \wedge (dg)^q$, where the 1-form g is characterized by the formula

$$\nabla_X \nu = 2\pi g(X).\nu \tag{7.21}$$

Formula (7.21) shows that if the connection ∇ in $\wedge^q Q^*$ preserves the volume form ν of Q, then $dg = 0$ and the Godbillon-Vey class is zero. If therefore the Bott connection in Q is volume preserving, the Godbillon-Vey class is zero. This can be seen directly from the fact that under the canonical map induced by

$$(GL(q),SO(q)) \longrightarrow (GL(q),SL(q))$$

the Godbillon-Vey class comes already from an element

$$[y_1 \otimes c_1^q] \in H^{2q+1}(W(\underline{\underline{gl}}(q),SL(q))_q).$$

The non-triviality of $\Delta_*(F(Q^*))(y_1 \otimes c_1^q) \in H_{DR}^{2q+1}(M)$ obstructs then by theorem 4.52 the Bott connection in the $GL(q)$-frame bundle $F(Q^*)$ from being induced by a foliation of the $SL(q)$-frame bundle of Q^*.

To give the original interpretation of the 1-form g by Godbillon-Vey [GV], observe first that under the orientability assumption on Q^* there exists locally a framing of Q^* by linearly independent 1-forms $\omega_1,\ldots,\omega_q$ such that

$$\nu = \omega_1 \wedge \ldots \wedge \omega_q \tag{7.22}$$

for the section ν of $\wedge^q Q^*$. By the theorem of Frobenius there exist then local 1-forms α_{ij} such that

$$d\omega_j = \sum_k \alpha_{jk} \wedge \omega_k \tag{7.23}$$

The connection in Q^* defined by the local formulae

$$\nabla_X^* \omega_j = \sum_k \alpha_{jk}(X)\omega_k$$

extends by (2.18) and (7.23) the Bott connection. For the induced connection in $\wedge^q Q^*$ we have then

$$\nabla^*_X \nu = \sum_j \omega_1 \wedge \ldots \wedge \nabla^*_X \omega_j \wedge \ldots \wedge \omega_q$$

$$= \sum_j \omega_1 \wedge \ldots \wedge i(X)\alpha_{jj} \wedge \omega_j) \wedge \ldots \wedge \omega_q = (\sum_j i(X)\alpha_{jj}) . \nu$$

so that by (7.21)

$$(7.24) \qquad 2\pi g = \sum_j \alpha_{jj}.$$

But note that by (7.22) (7.23) also

$$d\nu = \sum_j (-1)^{j-1} \omega_1 \wedge \ldots \wedge d\omega_j \wedge \ldots \wedge \omega_q = (\sum_j \alpha_{jj}) \wedge \nu$$

so that

$$(7.25) \qquad d\nu = 2\pi g \wedge \nu.$$

With this interpretation of g the original definition of the Godbillon-Vey invariant as the cohomology class of $g \wedge dg^q$ has an immediate appeal.

We observe further that the interpretation (7.21) of $2\pi g$ as a connection form in $\wedge^q Q^*$ leads immediately to the interpretation of $2\pi\, dg$ as the curvature form of this connection. It suffices to verify that (7.21) implies

$$(7.26) \qquad ([\nabla^*_X, \nabla^*_Y] - \nabla^*_{[X,Y]})\nu = 2\pi\, dg(X,Y) . \nu$$

which is immediate. Thus for a codimension 1 foliation the Godbillon-Vey class is (up to a factor) the exterior product $g \wedge v$ of the connection form g in Q^* with the curvature form $v = dg$. If in particular the foliation is defined as the kernel of a closed 1-form, then $g \wedge v$ is already zero as a form.

7.27 ROUSSARIES EXAMPLE. This is the first example of a non-trivial characteristic invariant for foliations given in [GV]. In $G = SL(2)$ consider the subgroup $H = ST(2)$ of triangular matrices $\binom{a\ 0}{b\ c}$ with $ac = 1$. The foliation of G by the left cosets of H is a codimension 1 foliation. The Godbillon-Vey form $g \wedge v$ is of degree 3. Let $\Gamma \subset SL(2)$ be a discrete uniform subgroup. Then the compact quotient $\Gamma\backslash SL(2)$ inherits the foliation of $SL(2)$. Since $g \wedge v$ is in fact invariant under the left $SL(2)$-action, it is the lift of a form on $\Gamma\backslash SL(2)$ which is the Godbillon-Vey-form of the foliation on $\Gamma\backslash SL(2)$. The idea is to verify that $g \wedge v$ is a volume form on $SL(2)$, (i.e. nowhere zero). Then it is also a volume form on $\Gamma\backslash SL(2)$, and hence $(g \wedge v)[M] \neq 0$.

The matrices

$$X = \begin{pmatrix} 1 & 0 \\ 0 & -1 \end{pmatrix}, \qquad Y = \begin{pmatrix} 0 & 0 \\ 1 & 0 \end{pmatrix}, \qquad Z = \begin{pmatrix} 0 & 1 \\ 0 & 0 \end{pmatrix}$$

are a basis of $\underline{\underline{g}}$, and X,Y a basis of $\underline{\underline{h}}$. The left invariant vectorfield Z is a section of the normal bundle Q of the left coset foliation of G by H. Let X^*,Y^*,Z^* be the dual basis of $\underline{\underline{g}}^*$. Then an explicit calculation (see e.g. [B 3], p. 62-64) shows that

$$g = -\frac{1}{\pi} X^*, \quad v = dg = -\frac{1}{\pi} Y^* \wedge Z^*$$

so that

$$g \wedge v = \frac{1}{\pi^2} X^* \wedge Y^* \wedge Z^*$$

which is a volume form on $SL(2)$. This establishes the non-triviality of the Godbillon-Vey invariant for the foliation on $\Gamma\backslash SL(2)$ induced by the foliation of $SL(2)$ by the left cosets of $ST(2)$.

For special choices of Γ the space $\Gamma\backslash SL(2)$ has the following interpretation. Let Γ_g be the group with $2g$ generators $\alpha_1,\ldots,\alpha_{2g}$ $(g > 1)$ and relation

$$\alpha_1\alpha_2\alpha_1^{-1}\alpha_2^{-1} \cdots \alpha_{2g-1}\alpha_{2g}\alpha_{2g-1}^{-1}\alpha_{2g} = 1.$$

It is well-known that $\Gamma_g \subset SL(2)$ and that $X_g = \Gamma_g\backslash SL(2)/SO(2)$ is a Riemannian surface of genus g. Then $\Gamma_g\backslash SL(2)$ is the unit tangent bundle of X_g.

7.28 REMARKS. The particular interest of Roussaries example comes from the fact that a discrete uniform subgroup $\Gamma \subset SL(2)$ can be deformed so that the volume of $\Gamma\backslash SL(2)$ varies continuously. It is then clear that for such a continuous family $M_t = \Gamma_t\backslash SL(2)$ the Godbillon-Vey number $(g \wedge v)[M_t]$ will take continuously varying values.

Thurston has shown in [Th 1] that there exist foliations of codimension 1 on S^3 such that the Godbillon-Vey number takes any prescribed real value. For the particular Reeb foliation on S^3 [R] the Godbillon-Vey number is zero. For generalizations of Roussaries example see 7.54 and 7.95 below.

7.29 RIEMANNIAN FOLIATIONS. As explained in detail in chapter 4, for the normal bundle Q of a Riemannian foliation of codimension q, the generalized characteristic homomorphism

$$\Delta(Q)_*: H(W(\underline{\underline{so}}(g),H)_{[q/2]}) \longrightarrow H_{DR}(M)$$

is defined for any reduction of the orthogonal frame bundle of Q to a subgroup $H \subset O(q)$. The map $\Delta_*(Q)$ has been computed for the case of a foliation of a group $\overline{G}$ by the left cosets of a subgroup G (see [KT 9,10] and also [LP]). This situation is discussed more generally in the following framework.

7.30 HOMOGENEOUS FOLIATIONS. We consider as in (2.44) (2.50) the foliated G-bundle

$$(7.31) \qquad P = (\Gamma\backslash\overline{G}) \times_H G \longrightarrow M = \Gamma\backslash\overline{G}/H.$$

Again $H \subset G \subset \overline{G}$ are subgroups of the Lie group $\overline{G}$ with H closed in $\overline{G}$, and $\Gamma \subset \overline{G}$ is a discrete subgroup operating properly discontinuously and without fixed points on $\overline{G}/H$, so that M is a manifold. The foliation of $\overline{G}$ by the left cosets of G induces a foliation of M with normal bundle Q_G of typical fibre $\overline{\underline{\underline{g}}}/\underline{\underline{g}}$. The bundle Q_G is associated to P and its characteristic classes (as foliated bundle) determined by those of P. To be precise, there is a commutative diagram

$$(7.32) \qquad \begin{array}{ccc} H(W(\underline{\underline{gl}}(\underline{\underline{m}}), O(\underline{\underline{m}}))_q) & \xrightarrow{\Delta_*(Q_G)} & \\ \Big\downarrow W(\rho^*) & & H_{DR}(M) \\ H(W(\underline{\underline{g}},H)_q) & \xrightarrow{\Delta_*(P)} & \end{array}$$

Here $\underline{\underline{m}} = \overline{\underline{\underline{g}}}/\underline{\underline{g}}$, $g = \dim \underline{\underline{m}}$ and $\rho : \underline{\underline{g}} \longrightarrow \underline{\underline{gl}}(\underline{\underline{m}})$ denotes the adjoint representation of $\underline{\underline{g}}$ in $\underline{\underline{m}}$ which associates Q_G to P. We assume that ρ represents the subgroup H in $O(q)$, which is certainly the case if e.g. H is compact. If the exact sequence $0 \longrightarrow \underline{\underline{g}} \longrightarrow \overline{\underline{\underline{g}}} \longrightarrow \overline{\underline{\underline{g}}}/\underline{\underline{g}} \longrightarrow 0$ has a G-equivariant splitting $\theta : \overline{\underline{\underline{g}}} \longrightarrow \underline{\underline{g}}$, we are in the basic situation and the truncation index q in (7.32) can be replaced by $[\frac{q}{2}]$.

The evaluation of $\Delta_*(P)$ is based on the following observation. For notational simplicity we assume in the following statement that $\Gamma = \{e\}$. Then $\overline{G}$ acts from the left on P. In the presence of a non-trivial Γ the connections considered on P

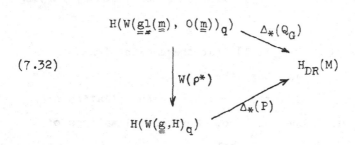

are locally $\overline{G}$-invariant. We have then the following result.

7.33 PROPOSITION [KT 10, 2.10]. *Let P be the canonically foliated homogeneous bundle (7.31). There is a bijection between left $\overline{G}$-invariant and adapted connections ω on P and H-equivariant splittings $\theta : \overline{\mathfrak{g}} \longrightarrow \mathfrak{g}$ of the exact sequence $0 \to \mathfrak{g} \to \overline{\mathfrak{g}} \to \overline{\mathfrak{g}}/\mathfrak{g} \to 0$, where ω and θ are related by the formula*

$$\omega_{(\overline{g},g)}(X_L, Y_R) = \mathrm{Ad}(g^{-1}) \circ (\theta(x) + y) \tag{7.34}$$

The evaluation of the cochain map

$$\Delta(\omega) : W(\mathfrak{g}, H)_q \longrightarrow \Omega(M)$$

realizing $\Delta_*(P)$ (see chapter 4) can therefore be done in terms of θ. This is the content of the following statement.

7.35 THEOREM [KT 9] [KT 10, 3.7]. *Let $H \subset G \subset \overline{G}$ be Lie groups, H closed in $\overline{G}$ with finitely many connected components, and $\Gamma \subset \overline{G}$ a discrete subgroup acting properly discontinuously and without fixed points on $\overline{G}/H$.*

The canonical G-foliation L_G of $M = \Gamma\backslash\overline{G}/H$ has codimension $q = \dim \overline{\mathfrak{g}}/\mathfrak{g}$. The G-bundle $P = (\Gamma\backslash\overline{G}) \times_H G \longrightarrow M$ is canonically foliated. Let ω be a locally $\overline{G}$-invariant adapted connection on P, characterized by an H-equivariant splitting θ of the exact sequence

$$0 \longrightarrow \mathfrak{g} \longrightarrow \overline{\mathfrak{g}} \longrightarrow \overline{\mathfrak{g}}/\mathfrak{g} \longrightarrow 0 \tag{7.36}$$

Then the generalized characteristic homomorphism $\Delta(\omega)$ of P on the cochain level factorizes as follows

$$(7.37)\qquad \begin{array}{ccc} W(\underline{\underline{g}},H)_q & \xrightarrow{\Delta(\theta)_H} & \Lambda^{\cdot}(\overline{\underline{\underline{g}}}/h)^{*H} \\ & \searrow^{\Delta(P)} & \downarrow \gamma \\ & & \Gamma(M,\Omega^{\cdot}_M) \end{array}$$

where γ is the canonical inclusion and $\Delta(\theta)_H$ is induced by the H-DG-homomorphism

$$\Delta(\theta) : W(\underline{\underline{g}}) \longrightarrow \Lambda\overline{\underline{\underline{g}}}^*$$

which is completely determined by

$$(7.37)\qquad \begin{aligned} \Delta(\theta)\alpha &= \alpha\theta \quad \text{for} \quad a \in \Lambda^1\underline{\underline{g}}^* \\ \Delta(\theta)\tilde{\alpha} &= d\alpha\theta + \tfrac{1}{2}\,\alpha[\theta,\theta] \quad \text{for} \quad \tilde{\alpha} \in S^1\underline{\underline{g}}^*. \end{aligned}$$

Under a stronger assumption on θ we have the following result.

7.39 THEOREM [CT 9] [KT 10, 3.11]. Let the situation be as in Theorem 7.35. Assume that θ is a G-equivariant splitting of the exact sequence (7.36). Then there exists an L_G-basic and locally $\overline{G}$-invariant adapted connection ω on P, and $\Delta(\omega)$ on the cochain level factorizes as follows

$$(7.40)\qquad \begin{array}{ccc} W(\underline{\underline{g}},H)_{[q/2]} & \xrightarrow{\Delta(\theta)} & \Lambda^{\cdot}(\overline{\underline{\underline{g}}}/\underline{\underline{h}})^{*H} \\ & \searrow^{\Delta(P)} & \downarrow \gamma \\ & & \Gamma(M,\Omega^{\cdot}_M) \end{array}$$

where γ, $\Delta(\theta)$ are defined as in theorem 7.35.

The point of these results is that $\Delta(\omega)$ appears as the composition of the map $\Delta(\theta)$, which is of a purely algebraic nature, and the map γ, which is the characteristic map of the flat bundle

(7.41) $$\Gamma\backslash\overline{G} \times_H \overline{G} \cong \overline{G}/H \times_\Gamma \overline{G} \longrightarrow \Gamma\backslash\overline{G}/H.$$

7.42 COSET FOLIATIONS FOR REDUCTIVE PAIRS $(\overline{G},G)$. Explicit computations lead to the following result.

7.43 THEOREM [KT 9] [KT 10, 6.49]. _Let_ $Q_{U(r)}$ _be the normal bundle of the foliation of_ $SU(r+1)$ _by the left cosets of the unitary group_ $U(r)$ _with quotient the complex projective space_ $\mathbb{P}^r\mathbb{C}$. _The image of the generalized characteristic homomorphism_ $\Delta_*(Q_{U(r)})$ _in_ $H^+(SU(r+1)) \cong \Lambda^+(\overline{y}_2, \ldots, \overline{y}_{r+1})$ _is the ideal generated by the primitive element_ $\overline{y}_{r+1}$, _the suspension of the top-dimensional Chern class_ $\overline{c}_{r+1} \in I^{2r+2}(SU(r+1))$.

This implies that $\dim \operatorname{im} \Delta_*^+(Q_{U(r)}) = 2^{r-1}$, whereas $\dim H(SU(r+1)) = 2^r$. It further shows the abundant existence of non-trivial linearly independent secondary invariants in dimensions greater than $2r+1$.

7.44 THEOREM [KT 9] [KT 10, 6.52]. _Let_ $Q_{SO(2r)}$ _be the normal bundle of the foliation of_ $SO(2r+1)$ _by the left cosets of the orthogonal group_ $SO(2r)$ _with quotient the sphere_ S^{2r}. _The image of_ $\Delta_*(Q_{SO(2r)})$ _in_ $H(SO(2r+1)) = \Lambda(\hat{y}_1,\ldots,\hat{y}_r)$ _is the direct sum_

$$\mathrm{Id}(\overline{y}_r) \oplus \Lambda(\overline{y}_{[r/2]+1}, \ldots, \overline{y}_{r-1})$$

of the ideal generated by the suspension $\overline{y}_r$ of the top-dimensional Pontrjagin class $\overline{p}_r \in I^{4r}(SO(2r+1))$, and the exterior algebra generated by the primitive elements $\bar{y}_{[r/2]+1}, \ldots, \bar{y}_{r-1}$.

It follows in particular that

$$(7.45) \qquad \dim \operatorname{im} \Delta_*(Q_{SO(2r)}) = \begin{cases} 2^{r-1} + 2^{[\frac{r}{2}]-1} & \text{for } r = 2k \\ 2^{r-1} + 2^{[\frac{r}{2}]} & \text{for } r = 2k+1 \end{cases}$$

These remarks are obtained from an analysis of $\Delta(\theta)$ for reductive pairs and more particularly symmetric pairs $(\bar{G},G)$. For this purpose the relevant complexes have to be replaced by the cohomology equivalent A-complexes according to the algorithm described in Chapter 5.

A complete determination of $\Delta(\theta)_*$ can be carried out for symmetric pairs $(\bar{G},G)$ of equal rank satisfying the following two conditions:

(7.46) the generators of $I(G) \cong \mathbb{R}[c_1,\ldots,c_r]$

can be chosen such that $\deg c_i < \deg c_k$ for $1 \leq i < k \leq r$;

(7.47) for $I(\bar{G}) \cong \mathbb{R}[\bar{c}_1, \ldots, \bar{c}_r]$ there exists $\bar{c}_j$ such that

$$i^*\bar{c}_j = \sum_{k=1}^{r} c_k \cdot \Phi_{jk}(c_k,\ldots,c_r)$$

with $\deg \Phi_{jk} = 2q' = \dim \bar{\underline{g}}/\underline{g}$ for the non-zero Φ_{jk} and $k = 1,\ldots,r$.

Theorems 7.43 and 7.44 are then consequences of the following result.

7.48 THEOREM [KT 10, 6.40]. _Let_ $(\bar{G},G)$ _be a symmetric pair of equal rank_ r _and satisfying conditions_ (7.46) _and_ (7.47). _Then for the generalized characteristic homomorphism_ $\Delta(\theta)_*$ _of the left coset foliation of_ $\bar{G}$ _by_ G _we have_

$$\operatorname{im} \Delta(\theta)_* = \operatorname{Id}(\bar{y}_j) + \wedge(y_{t+1},\ldots,y_r) \subset H(\underline{\underline{\bar{g}}})$$

The primitive class $\bar{y}_j$ *of* $\underline{\underline{\bar{g}}}$ *is the suspension of the distinguished generator* $\bar{c}_j$ *of* $I(\bar{G})$ *in* (7.47) *and the primitive classes* y_i *of* $\underline{\underline{g}}$ $(i = t+1,\ldots,r)$ *are the suspensions of the generators* c_i *of* $I(G)$ *satisfying* $\deg c_i > 2q' = \dim \underline{\underline{\bar{g}}}/\underline{\underline{g}}$.

The ideal $\operatorname{Id}(\bar{y}_j)$ generated by the element $\bar{y}_j$ has already dimension 2^{r-1}, since $H(\underline{\underline{\bar{g}}}) = \wedge(\bar{y}_1,\ldots,\bar{y}_r)$ has dimension 2^r. This produces linearly independent secondary invariants in dimensions greater than $2_{q'}+1$.

7.49 COSET FOLIATIONS FOR NON-REDUCTIVE PAIRS. We discuss the foliation of $\bar{G} = SL(r+1)$ by the subgroup $G = ST(r+1)$ of triangular matrices with determinant 1, and the induced foliation on the compact quotient by a discrete uniform subgroup $\Gamma \subset SL(r+1)$. The results are valid over the fields $\mathbb{R}$ or $\mathbb{C}$.

7.50 PROPOSITION [KT 9]. *The principal bundle*

$$P = \Gamma\backslash SL(r+1) \times ST(r+1) \longrightarrow M = \Gamma\backslash SL(r+1)$$

carries a canonical foliation compatible with the canonical foliation in the frame bundle $F(Q_G) = \Gamma\backslash SL(r+1) \times GL(V)$ *of the transversal bundle* Q_G, $V = \underline{\underline{s\ell}}(r+1)/\underline{\underline{st}}(r+1)$. *Moreover the adjoint representation* $\rho : ST(r+1) \longrightarrow GL(V)$ *realizes* P *as a subbundle of the flag bundle* $\operatorname{Flag}(Q_G)$ *in the frame bundle* $F(Q_G)$.

It follows that we have a commutative diagram

(7.51)
$$\begin{array}{ccc}
H(W(\underline{\underline{st}}(r+1)_q) & \xrightarrow{\Delta(\theta)_*} & H(\underline{\underline{s\ell}}(r+1)) \\
\uparrow{\scriptstyle W(\rho)^*} & \searrow{\scriptstyle \Delta(P)_*} & \downarrow{\scriptstyle \gamma_*} \\
H(W(\underline{\underline{st}}(V))_q) & \xrightarrow{\Delta(\operatorname{Flag})Q_G))_*} & H(\Gamma\backslash SL(r+1)
\end{array}$$

where $q = \dim V$.

In this case also one can use the techniques of chapter 5 to establish the following result.

7.52 THEOREM [KT 9]. Consider the cycle

$$z = \alpha_1 \wedge \ldots \wedge \alpha_r \otimes \tilde{\beta}^q \in W^n(\underline{\underline{st}}(r+1))_q,$$

where the $\alpha_i \in \Lambda^1 \underline{\underline{st}}$ are a basis and $\tilde{\beta} = \sum_{i=1}^{r} a_i \lambda_i \in I^2(\underline{\underline{st}}(r+1))$ with λ_i the i-th diagonal function on $\underline{\underline{st}}(r+1)$. Then

$$(7.53) \qquad \Delta(P)_*(z) = q! \prod_{1 \leq j < i \leq r+1} (a_j - a_i) \cdot \gamma_* \nu \in H^n(\Gamma \backslash SL(r+1))$$

where ν is an invariant volume on $\underline{\underline{s\ell}}(r+1)$ and $a_{r+1} = 0$, $n = 4(r+2)$ and $q = \binom{r+1}{2}$.

Since γ_* is injective, $\gamma_*\nu$ is a volume on $\Gamma \backslash SL(r+1)$. for $a_i \neq a_j$ $(i \neq j)$, $a_{r+1} = 0$ it follows that $\Delta(P)_*(z)$ is a non-trivial cohomology class in top degree of $\Gamma \backslash SL(r+1)$.

An example is $\beta = \operatorname{tr} \rho$, for which we obtain the non-trivial class

$$(7.54) \; \Delta(P)_*(\alpha_1 \wedge \ldots \wedge \alpha_r \otimes (\operatorname{tr} \tilde{\rho})^q) = 2^q . q! \prod_{1 \leq j < i \leq r+1} (i-j) . \gamma_* \nu \in H^n(\Gamma \backslash SL(r+1)).$$

Using this formula and the weights of the adjoint representation it is possible to show that $\Delta(\mathrm{Flag}(Q_G))_*$ is non-trivial. For $r = 1$ and $\alpha_1 = \operatorname{tr} \rho$ the class (7.54) is the Godbillon-Vey class in Roussaries example 7.27.

7.55 COSET FOLIATION BY MAXIMAL TORUS. By the same methods one computes the characteristic homomorphism for a reductive pair $(\bar{\underline{\underline{g}}}, \underline{\underline{g}})$, where $\underline{\underline{g}}$ is a Cartan algebra in the semi-simple Lie algebra $\bar{\underline{g}}$ over $\mathbb{C}$. The choice of a special basis in $\bar{\underline{g}}$ allows to compute $\Delta(\theta)$ explicitly.

Let $\{\bar{h}_{\alpha_i}, e_{\pm\alpha}\}$, $\alpha_i \in \Delta$, $\alpha \in \Phi^+$ be a Chevalley basis of $\bar{\underline{g}}$ ([BO 2], p. 115). Here Φ^+ denotes the set of positive roots and Δ the set of simple roots. $\bar{h}_\alpha$, $\alpha \in \Phi^+$ is defined by the Killing form of $\bar{\underline{g}}$ restricted to a rational form $\underline{g}_{\mathbb{Q}}$ of $\underline{g}$ by the formula

$$(\bar{h}_\alpha, x) = \frac{2}{(\alpha,\alpha)} \cdot \alpha(x) \quad \text{for} \quad x \in \underline{g}_{\mathbb{Q}}.$$

A $\underline{g}$-equivariant splitting θ of $0 \to \underline{g} \to \bar{\underline{g}} \to \bar{\underline{g}}/\underline{g} \to 0$ is defined by

$$\theta(h_{\alpha_i}) = h_{\alpha_i}, \quad \theta(e_{\pm\alpha}) = 0.$$

One has then the following result.

7.56 THEOREM [KT 9]. Let

$$z = \alpha_{i_1} \wedge \dots \wedge \alpha_{i_s} \otimes \tilde{\beta}^{q'} \in W(\underline{g})_{q'}$$

be a cocycle, where q' equals the number of positive roots and ρ is a weight $\in \underline{g}^*_{\mathbb{Q}}$. Then one has the formula

$$(7.57)\quad \Delta(\theta)z = q'!2^{q'} \prod_{\alpha\in\Phi^+} \frac{(\rho,\alpha)}{(\alpha,\alpha)} x^*_{i_1} \wedge \dots \wedge x^*_{i_s} \wedge \mu \in \Lambda\underline{g}^*,$$

where

$$\mu = \prod_{\alpha\in\Phi^+} (e^*_\alpha \wedge e^*_{-\alpha}) \in \Lambda^{2q'} V, \quad \bar{\underline{g}} = \underline{g} \oplus V$$

is the canonical volume of V and $x^*_i = \alpha_i \theta \in \bar{\underline{g}}^*$.

For $z = \alpha_1 \wedge \dots \wedge \alpha_r \otimes \tilde{\beta}$ one obtains in particular a class in the maximal dimension $n = \dim \bar{\underline{g}} = 2q' + 1$ of the form

$$(7.58) \qquad \Delta(\theta)z = q'!\, 2^{q'} \prod_{\alpha\in\Phi^+} \frac{(\rho,\alpha)}{(\alpha,\alpha)} \; v \in \Lambda^n \bar{\underline{\underline{g}}}^*,$$

where v is a volume of $\bar{\underline{\underline{g}}}$. Therefore $\Delta(\Theta)z \neq 0$ if one chooses $\rho \in \underline{\underline{g}}^*_{\mathbb{Q}}$ in the complement of the hyperplanes $(-,\alpha) = 0$ for $\alpha \in \Phi^+$.

Using the theory of real compact forms, it is then easy to obtain the characteristic homomorphism Δ_* for the foliation of a compact group by a maximal torus.

7.59 FOLIATED BUNDLES VERSUS FLAT BUNDLES [KT 10]. In this section we discuss a situation where the characteristic classes of a foliated bundle are to a large extent determined by the characteristic classes of a flat bundle. We begin with a flat $\bar{G}$-bundle $\bar{P} \longrightarrow X$. For any closed subgroup $G \subset \bar{G}$ the G-bundle $P = \bar{P} \longrightarrow P/G = M$ is foliated, as discussed repeatedly in (1.31), (2.21) and (2.40). There is a fibration $\hat{\pi} : M \longrightarrow X$ with fiber $\bar{G}/G$. In the following we assume that for the maximal compact subgroups $K_G \subset G$ and $K_{\bar{G}} \subset \bar{G}$ the canonical map

$$(7.60) \qquad K_{\bar{G}}/K_G \xrightarrow{\;\cong\;} \bar{G}/G$$

is a diffeomorphism. We wish then to compare the characteristic homomorphisms

$$(7.61) \qquad \begin{aligned} \Delta_*(P) &: H(W(\underline{\underline{g}},K_G)_q) \longrightarrow H_{DR}(M) \\ \Delta_*(\bar{P}) &: H(\bar{\underline{\underline{g}}},K_{\bar{G}}) \longrightarrow H_{DR}(X) \end{aligned}$$

of the foliated G-bundle $P \longrightarrow M$ and the flat $\bar{G}$-bundle $\bar{P} \longrightarrow X$.

Let Q_G denote the normal bundle of the foliation of M, which is defined by projecting the foliation on $\bar{P}$ giving its flat structure under the map $P = \bar{P} \longrightarrow M$. The integer q

occurring in (7.61) is the dimension of Q_G. Consider the exact sequence

$$(7.62)\qquad 0 \longrightarrow \underline{\underline{g}} \underset{\overset{\theta}{\longleftarrow}}{\longrightarrow} \overline{\underline{\underline{g}}} \longrightarrow \overline{\underline{\underline{g}}}/\underline{\underline{g}} = \underline{\underline{m}} \longrightarrow 0$$

it admits a K_G-invariant splitting θ, which composed with the flat connection form $\overline{\omega}$ on $\overline{P}$ gives an adapted connection $\omega = \theta \circ \overline{\omega}$ in P as in (2.40).

Let ρ denote the adjoint representation of G in $\underline{\underline{m}}$. Then $Q_G \cong T(\hat{\pi})$ is associated to P via ρ:

$$(7.63)\qquad Q_G \cong P \times_G \underline{\underline{m}}.$$

Let $s : M \longrightarrow P/K_G$ denote the (up to homotopy) canonical K_G-reduction of P and $\overline{s} : X \longrightarrow \overline{P}/K_G$ the $K_{\overline{G}}$-reduction of $\overline{P}$ uniquely determined by the commutative diagram

$$(7.64)\qquad \begin{array}{ccc} \overline{P}/K_{\overline{G}} & \xrightarrow{\ \overline{s}\ } & M \\ \downarrow{\scriptstyle \tilde{\pi}} & & \downarrow{\scriptstyle \hat{\pi}} \\ P/K_G & \xrightarrow{\ s\ } & X \end{array}$$

The desired relationship is then expressed on the cochain level by the commutative diagram

$$(7.65)\qquad \begin{array}{ccccc} W(\underline{\underline{gl}}(\underline{\underline{m}}),O(\underline{\underline{m}}))_q & & \xrightarrow{\qquad\qquad \Delta(Q_G) \qquad\qquad} & & \Omega(M) \\ \downarrow{\scriptstyle W(\rho)} & & \overset{\Delta(P,\omega,s)}{\curvearrowright} & & \| \\ W(\underline{\underline{g}},K_G)_q & \xrightarrow{k(\omega)K_G} & \Omega(P/K_G) & \xrightarrow{s^*} & \Omega(M) \\ \downarrow{\scriptstyle \Delta(\theta)} & & \| & & \| \\ (\wedge\overline{\underline{\underline{g}}}^*)_{K_G} & \xrightarrow{\overline{\omega}_{K_G}} & \Omega(\overline{P}/K_G) & \xrightarrow{s^*} & \Omega(M) \\ \downarrow{\scriptstyle j_*} & & \downarrow & & \downarrow{\scriptstyle \hat{\pi}_*} \\ (\wedge\overline{\underline{\underline{g}}}^*)_{K_G} & \xrightarrow{\overline{\omega}_{K_{\overline{G}}}} & \Omega(\overline{P}/K_{\overline{G}}) & \xrightarrow{\overline{s}^*} & \Omega(X) \\ & & \underset{\Delta(\overline{P},\overline{\omega},\overline{s})}{\smile} & & \end{array}$$

The map $j_* = i(\xi_1 \wedge \ldots \wedge \xi_q)$ denotes the interior product with the unique q-vector $\xi_1 \wedge \ldots \wedge \xi_q \in \wedge^q(\underline{\underline{k}}_{\overline{G}}/\underline{\underline{k}}_G)$ normalized by $i(\xi_1 \wedge \ldots \wedge \xi_q)\eta = 1$ for an invariant unit volume η on $K_{\overline{G}}/K_G$. If $j^* : (\wedge\underline{\underline{\overline{g}}}^*)_{K_{\overline{G}}} \longrightarrow (\wedge\underline{\underline{\overline{g}}}^*)_{K_G}$ denotes the canonical inclusion, then the derivation property of the interior product $i(\xi)$ leads to the following formula useful for computations:

$$j_*(j^*\alpha \cdot \beta) = \alpha \cdot j_*(\beta) \tag{7.66}$$

for a $\alpha \in (\wedge\underline{\underline{\overline{g}}}^*)_{K_{\overline{G}}}$, $\beta \in (\wedge\underline{\underline{\overline{g}}}^*)_{K_G}$.

The map $\hat{\pi}_*$ denotes integration over the fiber $K_{\overline{G}}/K_G \cong \overline{G}/G$ of the canonical map $\hat{\pi} : M \longrightarrow X$. The bottom rectangle in (7.65) is commutative (up to sign) (see [GHV], Vol. II, p. 243]).

The point of diagram (7.65) is that the vertical map on the left hand side is given completely in Lie algebra terms by the adjoint representation ρ of G in $\underline{\underline{m}}$, the split θ, and the canonical map j_*. If $\overline{P}$ is e.g. a locally homogeneous flat bundle as in theorem 4.87, then $\Delta(\overline{P})_* : H(\underline{\underline{\overline{g}}}, K_{\overline{G}}) \longrightarrow H_{DR}(X)$ is injective. Therefore the non-triviality of classes in the image of $\Delta(Q_G)_*$ surviving under $\hat{\pi}_*$ can be checked by purely Lie algebraic computations.

7.67 HOMOGENEOUS CASE. We apply this evaluation principle to the following locally homogeneous situation. Let $\overline{G}$ be a connected, non-compact and semi-simple group, and $\Gamma \subset \overline{G}$ a discrete uniform and torsion free subgroup. On the Clifford-Klein form $X = \Gamma\backslash\overline{G}/K_{\overline{G}}$ of the non-compact symmetric space $\overline{G}/K_{\overline{G}}$ there is then the flat $\overline{G}$-bundle

$$\overline{P} = (\Gamma\backslash\overline{G}) \times_{K_{\overline{G}}} \overline{G} \cong \overline{G}/K_{\overline{G}} \times_\Gamma \overline{G}. \tag{7.68}$$

Let $G \subset \overline{G}$ be a closed subgroup such that (7.60) holds. The locally homogeneous foliated G-bundle

$$P = \Gamma\backslash\overline{G} \times_{K_G} G \longrightarrow M = \Gamma\backslash\overline{G}/K_G \tag{7.69}$$

arises then from $\overline{P}$ in the way explained before.

To prove this we first describe the diffeomorphism in (7.68). Consider the map $\varphi : \overline{G} \times \overline{G} \longrightarrow \overline{G} \times \overline{G}$ defined by

$$\varphi(g,g') = (g,gg'). \tag{7.70}$$

We indicate its equivariance properties by the following diagram

$$G \times G \xrightarrow{\quad\varphi\quad} \overline{G} \times \overline{G} \tag{7.71}$$

There are three actions on each side of (7.71) given by:

$$^{\gamma}(g,g') = (\gamma g,g') \qquad\qquad \gamma.(g,g') = (\gamma g,\gamma g') \tag{7.72}$$

$$(g,g').\gamma = (g\gamma,\gamma^{-1}g') \qquad\qquad _{\gamma}(g,g') = (g\gamma,gg') \tag{7.73}$$

$$(g,g')^{\gamma} = (g,g'\gamma) \qquad\qquad (g,g')^{\gamma} = (g,g'\gamma) \tag{7.74}$$

The map φ is equivariant with respect to each pair of actions on the same line.

The equivariance corresponding to (7.73) implies for the subgroup $K_{\overline{G}} \subset \overline{G}$ (in fact for any closed subgroup) the diffeomorphism

$$\overline{G} \times_{K_{\overline{G}}} \overline{G} \cong \overline{G}/K_{\overline{G}} \times \overline{G} \tag{7.75}$$

The equivariances corresponding to (7.72) implies for $\Gamma \subset \overline{G}$ as above the diffeomorphism (7.68)

$$(\Gamma\backslash\overline{G}) \times_{K_{\overline{G}}} \overline{G} \cong \overline{G}/K_{\overline{G}} \times_{\Gamma} \overline{G}. \tag{7.76}$$

The RHS describes the flat bundle structure of $\overline{P} \longrightarrow X = \Gamma\backslash\overline{G}/K_{\overline{G}}$.

More generally φ induces for the smaller subgroup $K_G \subset G$ a map

$$(7.77) \qquad \overline{\varphi} : P = (\Gamma\backslash\overline{G}) \times_{K_G} G \longrightarrow \overline{P} = \overline{G}/K_{\overline{G}} \times_\Gamma \overline{G}$$

The equivariance corresponding to (7.74) implies that $\overline{\varphi}$ is G-equivariant with respect to the action on the second factor, i.e. a G-bundle map.

7.78 LEMMA. Under the assumption 7.60 the map $\overline{\varphi}$ induces on G-orbits a diffeomorphism

$$\overline{\varphi}_G : P/G \longrightarrow \overline{P}/G$$

and hence $\overline{\varphi} : P \longrightarrow \overline{P}$ is an isomorphism of G-bundles.

Proof. For P/G we have with obvious redundancies

$$(7.79) \qquad P/G = \Gamma\backslash\overline{G}/K_G \cong (\Gamma\backslash\overline{G}) \times_{K_G} G/G \cong (\Gamma\backslash\overline{G}) \times_{K_{\overline{G}}} K_{\overline{G}}/K_G$$

whereas

$$(7.80) \qquad \overline{P}/G = \overline{G}/K_{\overline{G}} \times_\Gamma \overline{G}/G \cong (\Gamma\backslash\overline{G}) \times_{K_{\overline{G}}} \overline{G}/G.$$

Since $\overline{\varphi}$ is induced from the identity on the first factor, so is $\overline{\varphi}_G$. On the second factor $\overline{\varphi}$ is induced from the canonical map $K_{\overline{G}}/K_G \longrightarrow \overline{G}/G$, which by assumption is a diffeomorphism. Therefore $\overline{\varphi}_G$ is a diffeomorphism. □

It follows that $M \cong \overline{P}/G$ and we are in the situation described before, i.e. the foliated bundle structure on the G-bundle P is induced from the flat $\overline{G}$-bundle structure on $\overline{P}$ by forming the quotient of $\overline{P}$ by G.

Note that the leaves of the foliation on

$$M = \Gamma\backslash\overline{G}/K_G \cong \overline{G}/K_{\overline{G}} \times_\Gamma \overline{G}/G$$

are transversal to $\overline{G}/G$ and each leaf projects as universal covering space onto X under $\hat{\pi} : M \longrightarrow X$.

For the normal bundle Q_G of the foliation on M we have then by (7.63)

(7.81) $$Q_G \cong P \times_G \underline{\underline{m}}$$

where $\rho : G \longrightarrow GL(\underline{\underline{m}})$ is the adjoint representation of G in $\underline{\underline{m}} = \overline{\underline{\underline{g}}}/\underline{\underline{g}}$. Further

(7.82) $$Q_G \cong T(\hat{\pi})$$

where $T(\hat{\pi})$ is the tangent bundle along the fiber of the canonical projection $\hat{\pi} : \overline{P}/G = M \longrightarrow \overline{P}/\overline{G} = X$. Our previous construction gives then the following result.

7.83 THEOREM [KT 10, 7.7]. Let $\overline{G}$ be a connected semi-simple Lie group with finite center and no compact factor, $K_{\overline{G}} \subset \overline{G}$ a maximal compact subgroup; and $\Gamma \subset \overline{G}$ a discrete, uniform and torsion free subgroup. Let $G \subset \overline{G}$ be a closed subgroup such that $K_{\overline{G}}/K_G \xrightarrow{\cong} \overline{G}/G$, where $K_G = K_{\overline{G}} \cap G$.

Let q be the codimension of the canonical G-foliation on $\Gamma\backslash\overline{G}/K_G$, with normal bundle $Q_G = T(\hat{\pi})$, $q = \dim \overline{\underline{\underline{g}}}/\underline{\underline{g}}$. Let $\theta : \overline{\underline{\underline{g}}} \longrightarrow \underline{\underline{g}}$ be a K_G-equivariant splitting of the exact sequence

(7.84) $$0 \longrightarrow \underline{\underline{g}} \longrightarrow \overline{\underline{\underline{g}}} \longrightarrow \overline{\underline{\underline{g}}}/\underline{\underline{g}} = \underline{\underline{m}} \longrightarrow 0$$

Then the generalized characteristic homomorphism $\Delta(Q_G)$ on the cochain level factorizes as in the following commutative diagram

(7.85)

$$\begin{array}{ccc}
W(\underline{\underline{gl}}(\underline{\underline{m}}),SO(\underline{\underline{m}}))_q & \overset{\Delta(Q_G)}{\searrow} & \\
\downarrow W(\rho) & & \\
W(\underline{\underline{g}},K_G)_q & \overset{\Delta(P)}{\longrightarrow} & \\
\downarrow \Delta(\theta)_{K_G} & & \\
(\Lambda\overline{\underline{\underline{g}}}^*)_{K_G} & \overset{\gamma}{\longrightarrow} & \Omega(\Gamma\backslash\overline{G}/K_G) \\
\downarrow j_* & & \downarrow \hat{\pi}_* \\
(\Lambda\overline{\underline{\underline{g}}}^*)_{K_{\overline{G}}} & \xrightarrow{\Delta(\overline{P}) = \overline{\gamma}} & \Omega(\Gamma\backslash\overline{G}/K_{\overline{G}}).
\end{array}$$

In this diagram, $\Delta(P)$ is the characteristic homomorphism of the foliated G-bundle P with its canonical K_G-reduction. γ denotes the canonical inclusion of the $\overline{G}$-invariant forms on $\overline{G}/K_G$ into the De Rham complex of $\Gamma\backslash\overline{G}/K_G$. $\Delta(\overline{P})$ is the characteristic homomorphism of the flat $\overline{G}$-bundle $\overline{P}$ in (7.68) with its canonical $K_{\overline{G}}$-reduction, namely the canonical inclusion $\overline{\gamma}$ of the $\overline{G}$-invariant forms on $G/K_{\overline{G}}$ into the De Rham complex of the Clifford-Klein form $\Gamma\backslash\overline{G}/K_{\overline{G}}$. It induces an injective cohomology map $\Delta(\overline{P})_*$ by Theorem 4.87.

In the following we explicitly compute $\Delta(Q_G)_*$ for pairs $(\overline{G},G)$ which typically are not reductive. This is in constrast to the examples discussed in 7.42. The pair $(\overline{G},K_{\overline{G}})$ is symmetric with $K_{\overline{G}}$-invariant decomposition

(7.86) $$\overline{\underline{\underline{g}}} = \underline{\underline{k}}_{\overline{G}} \oplus \overline{\underline{\underline{m}}}, \qquad [\overline{\underline{\underline{m}}},\overline{\underline{\underline{m}}}] \subset \underline{\underline{k}}_{\overline{G}}$$

given by the projection $\overline{\theta} : \overline{\underline{\underline{g}}} \longrightarrow \underline{k}_G$.

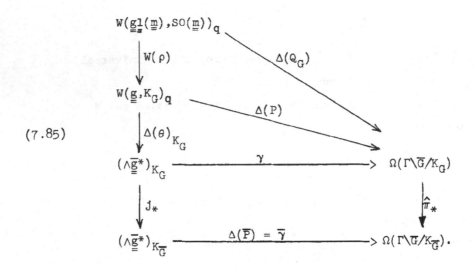

We use the notations

(7.87)

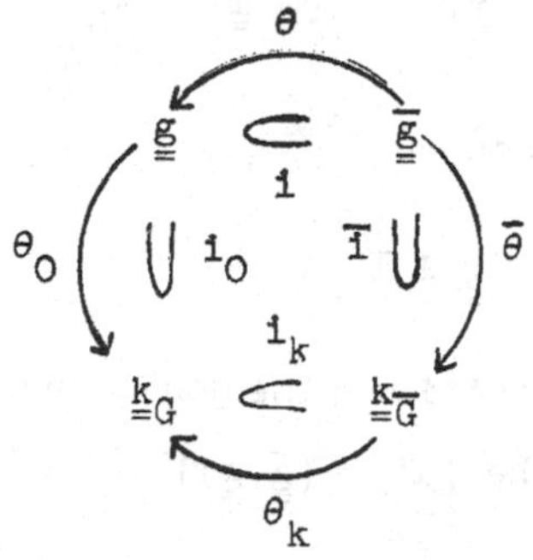

for the inclusion and splits of the various pairs. Note that the pairs $(\overline{\underline{\underline{g}}},\underline{\underline{g}})$ resp. $(\overline{\underline{\underline{g}}},\underline{\underline{k}}_{G})$, $(\underline{\underline{g}},\underline{\underline{k}}_{G})$ have all K_G-invariant splittings θ resp. $\theta_k \circ \overline{\theta}$, θ_0.

We assume in the sense of definition 5.101 that

(7.88) $\qquad (\overline{G},K_G)$ is a (CS)-pair.

This implies that the Samelson space $\hat{P}$ of $(\overline{\underline{\underline{g}}},\underline{\underline{k}}_{G})$ is given by

$$\hat{P}(\overline{\underline{\underline{g}}},\underline{\underline{k}}_{G}) = \overline{\sigma}(\ker \mathrm{res}),$$

where

$$\mathrm{res} = (\overline{i} \circ i_k)^* = (i \circ i_0)^* : I(\overline{G}) \longrightarrow I(K_G).$$

If $I(\overline{G}) = \mathbb{R}[\overline{c}_1,\ldots,\overline{c}_r]$ $(r = \operatorname{rank} \overline{\underline{\underline{g}}})$ and $P(\overline{\underline{\underline{g}}}) = \{\overline{y}_1,\ldots,\overline{y}_r\}$ for $\overline{y}_j = \overline{\sigma}\,\overline{c}_j$, then by renumbering the $\overline{c}_i$'s we may assume that

(7.89) $\qquad \hat{P}(\overline{\underline{\underline{g}}},\underline{\underline{k}}_{G}) = \overline{\sigma}\{\overline{c}_1,\ldots,\overline{c}_t\}$ for some $0 \leq t \leq r$.

We further assume that $(K_{\overline{G}}, K_G)$ is a symmetric pair with K_G-invariant decomposition

(7.90) $$\underline{\underline{k}}_{\overline{G}} = \underline{\underline{k}}_G \oplus \underline{\underline{m}}, \qquad [\underline{\underline{m}},\underline{\underline{m}}] \subset \underline{\underline{k}}_G$$

given by $\underline{\underline{m}} = \ker \theta_k$. Then

$$\overline{\underline{\underline{g}}} = \underline{\underline{g}} \oplus \overline{I}(\underline{\underline{m}})$$

is a K_G-invariant decomposition (the non G-invariance reflecting the non-reductivity of the pair $(\overline{\underline{\underline{g}}},\underline{\underline{g}})$).

According to the parity of $q = \dim \overline{\underline{\underline{g}}}/\underline{\underline{g}}$ we make then the following assumptions on the generators of $I(\overline{\underline{\underline{g}}})$:

(7.91) If $q = 2q'+1$, there exists a distinguished generator $\overline{c}_j$, $1 \leq j \leq t$ such that

$$i^*\overline{c}_j = \sum_k \Psi_{j,k} \cdot \Phi_{j,k} \neq 0 \quad \text{in} \quad I(G),$$

$$\overline{I}^*\overline{c}_j = \overline{c}'_{k_j} \cdot \Phi_j \neq 0 \quad \text{in} \quad I(K_{\overline{G}})$$

where $\Phi_{j,k} \in I(G)$ is of $\deg \Phi_{jk} = 2q$, $\Psi_{j,k} \in \ker(i_0^* : I(G) \to I(K_G))$ and $\overline{c}'_{k_j} \in I(K_{\overline{G}})$ is of $\deg \overline{c}'_{k_j} = q+1$, $\overline{\Phi}_j \notin \mathrm{Id}(\overline{I}^*I(\overline{G})^+)$. For the other generators $\overline{c}_\ell$, $1 \leq \ell \leq t$, $\ell \neq j$ we assume that

$$\overline{I}^*\overline{c}_\ell = 0 \quad \text{in} \quad I(K_{\overline{G}}).$$

(7.92) If $q = 2q'$, there exists a distinguished generator $\overline{c}_j$, $1 \leq j \leq t$ such that

$$i^*\overline{c}_j = \sum \Psi_{j,k}\, \Phi_{j,k} \neq 0 \quad \text{in} \quad I(G),$$

where $\Phi_{j,k} \in I(G)$ is of $\deg \Phi_{j,k} = 2q$, $\Psi_{j,k} \in \ker(i_0^* : I(G) \to I(K_G))$, and for all $\overline{c}_\ell$, $1 \leq \ell \leq t$

$$\overline{I^*c}_\ell = 0 \quad \text{in} \quad I(K_{\overline{G}})$$

In this case of even q, we further assume the symmetric pair $(K_{\overline{G}}, K_G)$ to be of equal rank. The Pfaffian polynomial $e_{q'} \in I^{2q'}(SO(\underline{\underline{m}}))$ gives rise to the tangent Euler class of $K_{\overline{G}}/K_G$ under the composition

$$I^{2q'}(SO(\underline{\underline{m}})) \xrightarrow{\rho^*} I^{2q'}(K_G) \xrightarrow{h(\theta_k)} (\Lambda^{2q'}\underline{\underline{m}}^*)^{K_G} \xrightarrow{\cong} H^{2q'}(K_{\overline{G}}/K_G)$$

Here $\rho : K_G \longrightarrow SO(\underline{\underline{m}})$ denotes the linear isotopy representation of K_G in $\underline{\underline{k}}_{\overline{G}}/\underline{\underline{k}}_G \cong \underline{\underline{m}}$. The map $h(\theta_k)$ is the Chern-Weil homomorphism of $K_G \longrightarrow K_{\overline{G}} \longrightarrow K_{\overline{G}}/K_G$ defined via the canonical splitting θ_k in 7.90. The differential in $(\Lambda\underline{\underline{m}}^*)^{K_G}$ is zero, so that $H(K_{\overline{G}}/K_G) \cong (\Lambda\underline{\underline{m}}^*)^{K_G}$. For an equal rank pair $(K_{\overline{G}}, K_G)$ the Euler number $\chi(K_{\overline{G}}/K_G)$ is well-known to be positive (it is the ratio of the order of the Weyl groups of $K_{\overline{G}}$ and K_G with respect to a common maximal torus).

7.93 THEOREM. Let the notations be as in theorem 7.83. Assume conditions (7.88) (7.90) and (7.91) (7.92) according to the parity of $q = \dim \overline{\underline{\underline{g}}}/\underline{\underline{g}} = \dim \underline{\underline{k}}_{\overline{G}}/\underline{\underline{k}}_G$. Then the following holds (see diagram (7.85)):

(i) $\Delta_*(\overline{P}) = \overline{\gamma}$ is injective;

(ii) γ_* is injective on the subspace spanned by the linearly independent cohomology classes of the 2^{t-1} cycles

$$\overline{y}_{i_1} \wedge \ldots \wedge \overline{y}_{i_s} \wedge \overline{y}_j \otimes 1, \quad \text{if} \quad q = 2q' + 1;$$

$$\overline{y}_{i_1} \wedge \ldots \wedge \overline{y}_{i_s} \wedge \overline{y}_j \otimes \rho^* e_{q'}, \text{if } q = 2q';$$

where $1 \leq i_1 < \ldots < i_s \leq t$, $i_\alpha \neq j$, $0 \leq s \leq t-1$;

(iii) <u>In fact we have for</u> $q = 2q' + 1$:

$$\hat{\pi}_* \gamma_* (\bar{y}_{i_1} \wedge \dots \wedge \bar{y}_{i_s} \wedge \bar{y}_j \otimes 1) = \kappa . \bar{\gamma}_* (\bar{y}_{i_1} \wedge \dots \wedge \bar{y}_{i_s} \otimes \Phi_j)$$

<u>where</u> $\bar{I}^* \bar{c}_j = \bar{c}'_{k_j} . \bar{\Phi}_j$ <u>and</u> $\kappa = \sigma_{K_{\bar{G}}}(\bar{c}'_{k_j})[K_{\bar{G}}/K_G] \neq 0$; <u>and for</u> $q = 2q'$

$$\hat{\pi}_* \gamma_* (\bar{y}_{i_1} \wedge \dots \wedge \bar{y}_{i_s} \wedge \bar{y}_j \otimes \rho^* e_{q'}) = \kappa . \bar{\gamma}_* (\bar{y}_{i_1} \wedge \dots \wedge \bar{y}_{i_s} \wedge \bar{y}_j \otimes 1)$$

where $\kappa = \chi(K_{\bar{G}}/K_G) > 0$;

(iv) $\operatorname{im} \Delta_*^+(Q_G) \subset H_{DR}^+(M)$ <u>contains the</u> γ_*<u>-image of the cohomology classes under</u> (ii), i.e.

$$\gamma_*(\bar{y}_j . \Lambda^{\cdot}(\bar{y}_1, \dots, \bar{y}_t) \otimes 1) \subset \Delta_*^+(Q_G) \quad \text{for } q = 2q + 1$$

$$\gamma_*(\bar{y}_j . \Lambda^{\cdot}(\bar{y}_1, \dots, \bar{y}_t) \otimes \rho^* e_{q'}) \subset \Lambda_*^+(Q_G) \quad \text{for } q = 2q'.$$

<u>In particular</u>

$$\dim \operatorname{im} \Delta_*^+(Q_G) \geq 2^{t-1}.$$

7.94 EXAMPLE. We apply these results to the groups $\bar{G} = SL(r+1)$ and $G = SL(r+1,1)_0$, the connected component of the group of unimodular matrices of the form $\left[\begin{array}{c|c} \lambda & * \\ \hline 0 & A \end{array}\right]$ with $A \in GL(r)$, $\det A = \lambda^{-1}$. The computation of the characteristic classes for this example has been carried out independently by Shulman and Tischler [ST]. In this case $K_{\bar{G}} = SO(r+1)$, $K_G = SO(r)$ and all previous assumptions hold. The kernel of the restriction

$$\begin{array}{ccc} I(SL(r+1)) & \longrightarrow & I(SO(r)) \\ \| \wr & & \cup \\ \mathbb{R}[\bar{c}_2, \dots, \bar{c}_{r+1}] & \longrightarrow & \mathbb{R}[p_1, \dots, p_{[\frac{r}{2}]}] \end{array}$$

is spanned by $\bar{c}_3, \bar{c}_5, \ldots, \bar{c}_{r'}, \bar{c}_{r+1}$, where $r' = 2[\frac{r+1}{2}] - 1$ is the largest odd integer $\leq r$. With $\bar{y}_j = \bar{\sigma}(\bar{c}_j) \in P_{\bar{\underline{\underline{g}}}}$, the Samelson space $\hat{P}(\bar{\underline{\underline{g}}}, \underline{\underline{k}}_G)$ is then spanned by the elements $\bar{y}_3, \bar{y}_5, \ldots, \bar{y}_{r'}, \bar{y}_{r+1}$. The distinguished generator occurring in conditions (7.91) (7.92) is $\bar{c}_{r+1}$ for both parities of r. Let $\Gamma \subset SL(r+1)$ be a discrete uniform and torsionfree subgroup. Then (7.93) implies the following result.

7.95 THEOREM. *Consider the spherical fibration*

$$M = \Gamma\backslash SL(r+1)/SO(r) \xrightarrow{\hat{\pi}} X = \Gamma\backslash SL(r+1)/SO(r+1)$$

over the Clifford-Klein form of the symmetric space $SL(r+1)/SO(r+1)$. *Then*

$$M \cong SL(r+1)/SO(r+1) \times_\Gamma SL(r+1)/SL(r+1,1)_0$$

carries a foliation of codimension $q = r$ *defined either by the right-action of* $SL(r+1,1)_0$ *on* $SL(r+1)$ *or by the flat structure of* M. *This foliation is transverse to the fiber* S^r *and every leaf is a universal covering space of* X *under the projection* $\hat{\pi}$. *The normal bundle* Q_G *of this foliation* $(G = SL(r+1,1)_0)$ *with its natural foliated structure is given by*

$$Q_G = T(\hat{\pi}) = \Gamma\backslash SL(r+1) \times_{SO(r)} \mathbb{R}^r \cong \Gamma\backslash(SL(r+1) \times_{SO(r)} SL(r+1,1)_0) \times_{SL(r+1,1)_0} \underline{\underline{m}}_\rho$$

where $\underline{\underline{m}}_\rho \cong \mathbb{R}^r$ *is equipped with the action* $\rho : SL(r+1,1)_0 \longrightarrow GL(r)$ *sending* $\left[\begin{array}{c|c} \lambda & * \\ \hline 0 & A \end{array}\right]$, $\lambda^{-1} = \det A$ *to* $\lambda^{-1}A$.

The characteristic homomorphism

$$\Delta(Q_G)_* : H(W(\underline{\underline{gl}}(r), SO(r))_r) \longrightarrow H_{DR}(M)$$

of this foliation has then the following properties (*see diagram* (7.85) *on* p. 176).

(i) <u>For</u> $2 \leq i_1 < \ldots < i_s \leq n$, $0 \leq s \leq n-1$; <u>and</u>

$$z_{(i,j)} = y_1 \wedge y_{2i_1-1} \wedge \ldots \wedge y_{2i_s-1} \otimes c_{(j)}, \quad c_{(j)} = c_1^{j_1} \ldots c_q^{j_q}$$

<u>with</u> $\deg c_{(j)} = 2\sum_{\ell=1}^{q} j_\ell \cdot \ell = 2q \ (q = r)$:

$$\Delta(Q_G)_*[z_{(i,j)} \otimes \Phi] = \mathscr{H} \cdot \gamma_* [\bar{y}_{2i_1-1} \wedge \ldots \wedge \bar{y}_{2i_s-1} \wedge \bar{y}_{q+1} \otimes \Phi],$$

<u>where</u> $\Phi = 1$ <u>for</u> $q = 2n-1$ <u>and</u> $\Phi = 1$ <u>or</u> e_n <u>for</u> $q = 2n$, <u>and</u> $\mathscr{H} = (-1)^{s+1}(q+1)\prod_{\ell=1}^{q} \binom{q+1}{\ell}^{j_\ell}$.

(ii) <u>For</u> $2 \leq i_0 \leq i_1 < \ldots < i_s \leq n$, $0 \leq s \leq n-1$; <u>and</u>

$z_{(i,j)} = y_{2i_0-1} \wedge \ldots \wedge y_{2i_s-1} \otimes c_{(j)}$,
$\deg c_{(j)} = 2(q+1-(2i_0-1))$:

$$\Delta(Q_G)_*[z_{(i,j)} \otimes \Phi] = \mathscr{H} \cdot \gamma_*[\bar{y}_{2i_1-1} \wedge \ldots \wedge \bar{y}_{2i_s-1} \wedge \bar{y}_{q+1} \otimes \Phi],$$

<u>where</u> Φ <u>as in</u> (i) <u>and</u> $\mathscr{H} = (-1)^{s+1}\prod_{\ell=1}^{q} \binom{q+1}{\ell}^{j_\ell}$.

<u>If</u> $\deg c_{(j)} > 2(q+1-(2i_0-1))$, <u>then</u>

$\Delta(Q_G)_* [z_{(i,j)} \otimes \Phi] = 0$.

<u>Composing</u> $\Delta(Q_G)_*$ <u>with the integration over the fiber map</u> $\hat{\pi}_*$ (<u>diagram</u> 7.85), <u>we have the following result</u>. <u>Let</u> $z_{(i,j)}$ <u>be as in</u> (i) <u>or</u> (ii).

(iii) <u>For</u> $q = 2n-1$

$$\hat{\pi}_* \Delta(Q_G)_*[z_{(i,j)}] = \mathscr{H} \cdot \bar{\gamma}_* j_* \Delta(\theta)_* [z_{(i,j)}]$$

$$= \mathscr{H} \cdot \langle \sigma e_n, S^{2n-1} \rangle \cdot \bar{\gamma}_*[\bar{y}_{2i_1-1} \wedge \ldots \wedge \bar{y}_{2i_s-1}] \otimes e,$$

<u>where</u> $\mathscr{H}$ <u>is as in</u> (i) (ii), $\bar{\gamma}_*$ <u>is injective by</u> 4.87 <u>and by the proof of</u> 6.49

$$<\sigma e_n, S^{2n-1}> = \frac{-1}{2^{2(n-1)}(2n-1)} .$$

(iv) For $q = 2n$

$$\hat{\pi}_* \Delta(Q_G)_*[z_{(i,j)} \otimes e_n] = \mathscr{H} . \bar{\gamma}_* j_* \Delta(\theta)_*[z_{(i,j)} \otimes e_n]$$

$$= \mathscr{H} . <e_n, S^{2n}> . \bar{\gamma}_*[\bar{y}_{2i_1-1} \wedge \dots \wedge \bar{y}_{2i_s-1} \wedge \bar{y}_{2n+1} \otimes 1]$$

where $\mathscr{H}$ is as in (i) or (ii), $<e_n, S^{2n}> = 2$ and $\bar{\gamma}_*$ is again injective.

This implies then the following facts.

(v) Let $q = 2n - 1$. Then

$\operatorname{im} \Delta(Q_G)_*^{+} = \gamma_*[\bar{y}_{2n} \cdot \wedge(\bar{y}_3, \dots, \bar{y}_{2n-1})]$,

γ^* is injective on the ideal $\bar{y}_{2n} \cdot \wedge(\bar{y}_3, \dots, \bar{y}_{2n-1})$ in $H(\underline{\underline{sl}}(2n), SO(2n-1))$, and $\Delta(Q_G)_*$ maps the cocycles

$$z_{(i,j)} = y_1 \wedge y_{2i_1-1} \wedge \dots \wedge y_{2i_s-1} \otimes c_q$$

with $2 \leq i_1 < \dots < i_s \leq n$, $0 \leq s \leq n-1$ onto an $\mathbb{R}$-basis of $\operatorname{im} \Delta(Q_G)_*^{+} \subset H_{DR}^{+}(M)$, namely (up to a real factor) the classes

$$\gamma_*[\bar{y}_{2i_1-1} \wedge \dots \wedge \bar{y}_{2i_s-1} \wedge \bar{y}_{2n}].$$

It follows that

$$\dim \operatorname{im} \Delta(Q_G)_*^{+} = 2^{n-1} .$$

(vi) Let $q = 2n$. Then

$$\operatorname{im} \Delta(Q_G)_*^{+} = \gamma_*[\bar{y}_{2n+1} \cdot \wedge(\bar{y}_3, \dots, \bar{y}_{2n-1}) \otimes \mathbb{R}[e_n]/(e_n^2)] ,$$

γ_* *is injective on the ideal*
$\bar{y}_{2n+1} \cdot \Lambda(\bar{y}_3, \ldots, \bar{y}_{2n-1}) \otimes e_n \subset H(\underline{\underline{sl}}(2n+1), SO(2n))$, *and*
$\Delta(Q_G)_*$ *maps the cocycles*

$$z_{(i,j)} \otimes e_n = y_1 \wedge y_{2i_1-1} \wedge \cdots \wedge y_{2i_s-1} \otimes c_q \otimes e_n$$

with $2 \leq i_1 < \ldots < i_s \leq n$, $0 \leq s \leq n-1$ *onto linearly independent elements of* $\operatorname{im} \Delta(Q_G)_*^+ \subset H_{DR}^+(M)$, *namely (up to a real factor) the classes*

$$\gamma_*[\bar{y}_{2i_1-1} \wedge \cdots \wedge \bar{y}_{2i_s-1} \wedge \bar{y}_{2n+1} \otimes e_n].$$

It follows that

$$\dim \operatorname{im} \Delta(Q_G)_*^+ \geq 2^{n-1}.$$

(*Note that for the cocycles* $z_{(i,j)} \otimes 1$, *integration over the fiber* S^{2n} *gives* 0 *as this class is already* $K_{\bar{G}} = SO(2n+1)$*-basic. No injectivity result about* γ_* *on these cocycles is known.*)

(vii) *The Pontrjagin ring of* Q_G *is trivial*:

$$p_i(Q_G) = 0 \quad \text{for} \quad i > 0.$$

For the special case $r = q = 1$, and $\Gamma_g \subset SL(2)$ the fundamental group of a Riemannian surface $X_g = \Gamma_g \backslash SL(2)/SO(2)$ of genus $g > 1$, the circle bundle

$$M_g = \Gamma_g \backslash SL(2) \xrightarrow{\hat{\pi}} X_g$$

is the unit tangent bundle of X_g. The foliation of M_g is induced by the left cosets of the triangular group $ST(2)$ in $SL(2)$ as discussed in 7.27. The nontrivial class

$$-4 \cdot \gamma_*(\bar{y}_2) \in H^3(M_g)$$

which according to (7.95) lies in $\operatorname{im} \Delta_*^+(Q_G)$, is in fact the negative of the image of the Godbillon-Vey class $y_1 \otimes c_1$ under $\Delta_*(Q_G) : H(W(\underline{\underline{gl}}(1))_1) \longrightarrow H_{DR}(M_g)$.

At this place we wish to point out that these linear independence result for secondary characteristic classes imply linear independence results in $H(B_{\Gamma_q^+})$, the cohomology of the classifying space for codimension q foliations with oriented normal bundle. Let indeed $f : M \longrightarrow B_{\Gamma_q^+}$ denote the classifying map for such a foliation on M. Then there is a commutative diagram

$$\begin{array}{ccc} & \xrightarrow{\Delta_*} & H(B_{\Gamma_q^+}) \\ H(W(\underline{\underline{gl}}(q), SO(q))_q) & & \downarrow f^* \\ & \xrightarrow{\Delta(Q_G)_*} & H_{DR}(M) \end{array}$$

where Δ_* is the generalized characteristic homomorphism of the universal codimension q-foliation on $B_{\Gamma_q^+}$. This is a consequence of the functoriality of the characteristic homomorphism (see 4.59). Cohomology classes which are realized in $H_{DR}(M)$ as linearly independent classes must a fortiori be linearly independent in $H(B_{\Gamma_q}^{\ +})$. Thus e.g. the classes given by the cocycles

$$y_1 \wedge y_{2i_1-1} \wedge \cdots \wedge y_{2i_s-1} \otimes c_q \otimes \Phi$$

for all $2 \leq i_1 < \ldots < i_s \leq n$, $0 \leq s \leq n-1$ and with $\Phi = 1$ for $q = 2n-1$ and $\Phi = e_n$ for $q = 2n$ are realized linearly independently in $H(B_{\Gamma_q^+})$ under the map Δ_*. The same holds true for the cohomology classes of the cocycles

$$y_{2i_0-1} \wedge \cdots \wedge y_{2i_s-1} \otimes c_p$$

for all $2 \leq i_0 < \ldots < i_s \leq n$, $0 \leq s \leq n-1$ and $2i_0 - 1 + p = q + 1$.

8. SEMI-SIMPLICIAL WEIL ALGEBRAS

8.1 OUTLINE. If one wants to apply the methods so far developed to complex analytic manifolds and algebraic varieties, one obstacle is the fact that no global connections need exist on holomorphic bundles or algebraic bundles P. This is a difficulty occurring already if one wants to use the ordinary Chern-Weil construction for P. But what one can use instead is a family of local connections on P restricted to the sets of a cohomologically trivial open covering of M. Even in the smooth case connections are often given in this way, and a direct construction of the characteristic classes of P via these data seems desirable, regardless of the fact that they can be constructed by using a global connection in P. If one wants to work with these data directly, one is then lead automatically to semi-simplicial methods, as the resulting invariants are defined via Čech cohomology. The basic idea for the construction of the generalized characteristic homomorphism $\Delta_*(P)$ in this situation is then as follows ([KT 6,7,8,12]). Let $\mathfrak{U} = (U_j)$ be a Γ-acyclic open covering of the base space of the foliated G-bundle $P \xrightarrow{\pi} M$. Then there exists a family $\omega = (\omega_j)$ of local connections ω_j on $P|U_j$, adapted to the flat partial connection. ω defines then a formal connection in the $\underline{\underline{g}}$-DG-algebra of Čech cochains $\check{C}^{\cdot}(\mathfrak{U}, \pi_*\Omega_P^{\cdot})$ of $\mathfrak{U}$ with coefficients in the direct image of the De Rham complex $\Omega_P^{\cdot}$ of P. A semi-simplicial model $W_1(\underline{\underline{g}})$ of the Weil algebra $W(\underline{\underline{g}})$ is constructed, on which ω defines a $\underline{\underline{g}}$-DG-algebra homomorphism

$$k_1(\omega) : W_1^{\cdot}(\underline{\underline{g}}) \longrightarrow \check{C}^{\cdot}(\mathfrak{U}, \pi_*\Omega_P^{\cdot}).$$

This generalized Weil homomorphism restricts then on H-basic elements for a subgroup H to the map giving rise to the characteristic homomorphism Δ_* of P. A more detailed description of this construction is as follows.

First one defines a sequence of semi-simplicial models $W_s(\underline{\underline{g}})$ for the Weil algebra $W(\underline{\underline{g}})$. The construction of $W_1(\underline{\underline{g}})$ is similar to the Amitsur complex of $W(\underline{\underline{g}})$. $W_s(\underline{\underline{g}})$ for $s > 1$ is obtained by an iterative process and $W_0(\underline{\underline{g}}) = W(\underline{\underline{g}})$. The essential feature of these algebras is that they behave cohomologically exactly like the Weil algebra. More precisely, there are even filtrations $F_s(\underline{\underline{g}})$ on $W_s(\underline{\underline{g}})$ for $s \geq 0$ and canonical filtration preserving homomorphisms $\rho_s : W_s(\underline{\underline{g}}) \longrightarrow W_{s-1}(\underline{\underline{g}})$ for $s > 0$ inducing cohomology-isomorphisms of the associated graded algebras and hence also of the truncated algebras $W_s(\underline{\underline{g}}) = W_s/F_s^{2(k+1)}$, $s \geq 0$, $0 \leq k \leq \infty$. The same is true for $\underline{\underline{h}}$-basic subalgebras provided $\underline{\underline{h}} \subset \underline{\underline{g}}$ satisfies certain conditions (theorem 8.12). The algebra $W_1(\underline{\underline{g}})$ is non-commutative except in trivial cases. Using this it is possible to show that the projection $\rho_1 : W_1(g) \longrightarrow W(\underline{\underline{g}})$ does not admit a splitting by a $\underline{\underline{g}}$-DG-algebra-homomorphism. However a splitting $\lambda : W(\underline{\underline{g}}) \longrightarrow W_1(\underline{\underline{g}})$ of ρ_1 by a $\underline{\underline{g}}$-DG-module map λ does exist. This is the map of theorem 5.35.

Then one constructs the homomorphism $k_1(\omega)$ mentioned above. It is filtration preserving, where the filtration on the target complex $\check{C}^\bullet(\mathcal{U},\pi_*\Omega_P^\bullet)$ is given by the ideals generated by the powers of Q^* in $\pi_*\Omega_P^\bullet$. For an H-structure of P defined by a cross-section $s : M \longrightarrow P/H$ the generalized characteristic homomorphism Δ_* is then also defined in this more general context. The same properties as in theorem 4.43 hold.

We finally describe explicitly some invariants associated to a foliated holomorphic vectorbundle whose first integral Chern class is zero.

8.2 SEMI-SIMPLICIAL WEIL ALGEBRAS. All algebras are defined over a fixed groundfield K of characteristic zero. From the description above we see that we have to consider two types of semi-simplicial $\underline{\underline{g}}$-DG-algebras, namely the algebras $W_s(\underline{\underline{g}})$ (to be defined below) and $\check{C}^{\cdot}(\mathcal{U},\pi_*\Omega^{\cdot}_p)$, and also homomorphisms between them. The common concept is the algebra $C^{\cdot}(S,\underset{\sim}{A})$ of cochains on a semi-simplicial set S with coefficients in a local system $\underset{\sim}{A}$ of (associative, but not necessarily commutative) $\underline{\underline{g}}$-DG-algebras on S.

Recall that a semi-simplicial set S is given by p-simplices S_p for $p \geq 0$, and face operators $\varepsilon^p_i : S_{p+1} \longrightarrow S_p$, degeneracy operators $\mu^p_i : S_p \longrightarrow S_{p+1}$, subject to the usual conditions under composition (see e.g. [GT], p. 271). A local system $\underset{\sim}{A}^{\cdot}$ of $\underline{\underline{g}}$-DG-algebras on S is the assignment of a graded $\underline{\underline{g}}$-DG-algebra $A^{\cdot}_\sigma$ for every $\sigma \in S_p$ and $\underline{\underline{g}}$-DG-homomorphisms

$$\overline{\varepsilon}^p_i(\sigma) : A^{\cdot}_{\varepsilon^p_i(\sigma)} \longrightarrow A^{\cdot}_\sigma, \quad \overline{\mu}^p_i(\sigma) : A^{\cdot}_{\mu^p_i(\sigma)} \longrightarrow A^{\cdot}_\sigma$$

with obvious composition rules. Local systems pull back canonically under maps of semi-simplicial sets. For S a semi-simplicial set and $\underset{\sim}{A}$ a local system of $\underline{\underline{g}}$-DG-algebras, the $\underset{\sim}{A}$-valued cochains on S

$$(8.3) \qquad C^{\cdot}(S,\underset{\sim}{A}) = \bigoplus_p C^p(S,\underset{\sim}{A}^{\cdot}), \quad C^p(S,\underset{\sim}{A}^{\cdot}) = \prod_{\sigma\in S_p} A^{\cdot}_\sigma$$

together with the maps $\varepsilon^p_i : C^p \longrightarrow C^{p+1}$, $\mu^p_i : C^{p+1} \longrightarrow C^p$ defined by

$$(\varepsilon^p_i c)(\sigma) = \overline{\varepsilon}^p_i \; c(\varepsilon^p_i\sigma), \quad (\mu^p_i c)(\sigma) = \overline{\mu}^p_i \, c(\mu^p_i\sigma)$$

is a (co-) semi-simplicial object in the category of $\underline{\underline{g}}$-DG-algebras. $C^{\cdot}(S,\underset{\sim}{A}^{\cdot})$ is functorial, covariant for maps of local systems $\underset{\sim}{A}^{\cdot} \longrightarrow \underset{\sim}{A}'^{\cdot}$ and contravariant for maps of semi-simplicial sets

$S \longrightarrow S'$.

The Alexander-Whitney multiplication in $C^\cdot(S,\underset{\sim}{A}^\cdot)$ is defined as follows. Denote $C^{p,q} = C^p(S,\underset{\sim}{A}^q)$. Then

$$(8.4) \qquad m_C : C^{p,q} \otimes C^{p',q'} \longrightarrow C^{p+p',q+q'}$$

is the composition $m_C = \mu_p^{p+p'}.(-1)^{p'q}\, m_A \circ \eta \otimes \zeta$, where

$$\eta = \varepsilon^{p+p'}_{p+p'+1} \circ \ldots \circ \varepsilon^p_{p+1}, \qquad \zeta = \varepsilon^{p+p'}_p \circ \ldots \circ \varepsilon^{p'}_0$$

and m_A is induced by the (associative) multiplication in $\underset{\sim}{A}$. This turns $C^\cdot(S,\underset{\sim}{A}^\cdot)$ into an associative graded algebra. It is non-commutative, even if $\underset{\sim}{A}$ is a local system of commutative $\underline{\underline{g}}$-DG-algebras. The differential d_A in $\underset{\sim}{A}$ and the semi-simplicial boundary operator $\delta = \sum_{i=0}^{p+1} (-1)^i \varepsilon_i^p$ turn $C^\cdot(S,\underset{\sim}{A}^\cdot)$ into a double complex with total differential $d_C = \delta + (-1)^p d_{\underset{\sim}{A}}$ on $C^{p,q}$. The complex $(C^\cdot(S,\underset{\sim}{A}^\cdot),d_C)$ is then a $\underline{\underline{g}}$-DG-algebra, where the operators $\theta(x)$, $i(x)$ are defined for each simplex, i.e.
$(i(x)c)(\sigma) = (-1)^p i(x)(c(\sigma)$, $(\theta(x)c)(\sigma) = \theta(x)(c(\sigma))$, $c \in C^{p,q}$, $x \in \underline{\underline{g}}$.

The example of interest in our geometric context is the case $S = N(\mathcal{U})$ for the nerve of an open covering $\mathcal{U} = (U_j)$ of M, and $\underset{\sim}{A} = \Gamma(-,\pi_*\Omega_P^\cdot)$ the local system defined by the direct image of the De Rham complex of P under $P \xrightarrow{\pi} M$. This is the algebra $\check{C}^\cdot(\mathcal{U},\pi_*\Omega_P^\cdot)$ discussed in the outline of this chapter. A family $\omega = (\omega_j)$ of local connections ω_j in $P|U_j$ adapted to the given partial flat connection on P defines then in the sense of definition 5.11 a "connection" $\wedge\underline{\underline{g}}^* \xrightarrow{\omega} \check{C}^o(\mathcal{U},\pi_*\Omega_P^1) \subset \check{C}^\cdot(\mathcal{U},\pi_*\Omega_P^\cdot)$. But since the target is a non-commutative $\underline{\underline{g}}$-DG-algebra, we cannot use the universal property of $W(\underline{\underline{g}})$ to extend ω to an algebra homomorphism on $W(\underline{\underline{g}})$. The remedy consists in applying the construction above once more to obtain a semi-simplicial algebra $W_1(\underline{\underline{g}})$ replacing $W(\underline{\underline{g}})$.

Here we return to the constructions in 5.26 and 5.34, which are better understood in the following context. Consider first a semi-simplicial object in the category of Lie algebras defined by $\underline{\underline{g}}$ as follows. Let $\underline{\underline{g}}^{\ell+1}$ denote for $\ell \geq 0$ the $(\ell+1)$-fold product of $\underline{\underline{g}}$ with itself. Define for $0 \leq i \leq \ell+1$, $0 \leq j \leq \ell$

$$\varepsilon_i^\ell : \underline{\underline{g}}^{\ell+2} \longrightarrow \underline{\underline{g}}^{\ell+1}, \quad \varepsilon_i^\ell(x_0,\ldots,x_{\ell+1}) = (x_0,\ldots,x_{i-1},x_{i+1},\ldots,x_{\ell+1})$$

$$\mu_j^\ell : \underline{\underline{g}}^{\ell+1} \longrightarrow \underline{\underline{g}}^{\ell+2}, \quad \mu_j^\ell(x_0,\ldots,x_\ell) = (x_0,\ldots,x_j,x_j,x_{j+1},\ldots,x_\ell).$$

Then ε and μ are the face and degeneracy maps for the semi-simplicial object in question and satisfy the usual relations.

Next consider the Weil-algebra as a contravariant functor from Lie algebras to $\underline{\underline{g}}$-DG-algebras and apply it to the semi-simplicial object discussed. This gives rise to a cosemi-simplicial object $W_1(\underline{\underline{g}})$ in the category of $\underline{\underline{g}}$-DG-algebras. Note that

$$W_1^\ell(\underline{\underline{g}}) = W(\underline{\underline{g}}^{\ell+1}) = W(\underline{\underline{g}})^{\otimes \ell+1}$$

and the face and degeneracy maps $\varepsilon_i^\ell = W(\varepsilon_1^\ell) : W_1^\ell \longrightarrow W_1^{\ell+1}$, $\mu_i^\ell = W(\mu_i^\ell) : W_1^{\ell+1} \longrightarrow W_1^\ell$ are given by the inclusions omitting the i-th factors and multiplication of the i-th and (i+1)-th factors. Thus $W_1(\underline{\underline{g}})$ is the Amitsur complex of $W(\underline{\underline{g}})$.

Let now Pt be the semi-simplicial point (terminal object) with exactly one simplex σ_ℓ in each dimension $\ell \geq 0$ and canonical face and degeneracy maps. As a local system on Pt is precisely given by a cosemi-simplicial object of $\underline{\underline{g}}$-DG-algebras, we may consider $W_1^{\cdot}(\underline{\underline{g}})$ as the cochain complex on Pt with coefficients in the local system $\underset{\sim}{W} : \sigma_\ell \longrightarrow W(\underline{\underline{g}}^{\ell+1})$, $\ell \geq 0$:

$W_1^{\cdot}(\underline{\underline{g}}) = C^{\cdot}(Pt, \underset{\sim}{W})$. As such $W_1(\underline{\underline{g}})$ is equipped with the Alexander-Whitney multiplication and thus carries canonically the structure of a $\underline{\underline{g}}$-DG-algebra. Observe that the $\underline{\underline{g}}$-actions on $W(\underline{\underline{g}}^{\ell+1})$ are induced by the diagonal $\Delta : \underline{\underline{g}} \longrightarrow \underline{\underline{g}}^{\ell+1}$. The construction performed with the functor W can now obviously be repeated with the functor W_1. By iteration we thus obtain a sequence of (co-)semi-simplicial $\underline{\underline{g}}$-DG-algebras $W_s(\underline{\underline{g}})$, $s > 0$, which will turn out to be proper substitutes for the commutative Weil-algebra $W(\underline{\underline{g}}) = W_0(\underline{\underline{g}})$. Note that by construction

(8.5) $$W_s(\underline{\underline{g}}) = \underset{\ell \geq 0}{\otimes} W_s^{\ell}(\underline{\underline{g}}) = \underset{\ell \geq 0}{\otimes} W_{s-1}(\underline{\underline{g}}^{\ell+1}).$$

The canonical projections

(8.6) $$\rho_s : W_s(\underline{\underline{g}}) \longrightarrow W_s^0(\underline{\underline{g}}) = W_{s-1}(\underline{\underline{g}}), \quad s > 0$$

are $\underline{\underline{g}}$-DG-algebra homomorphisms.

We proceed now to define inductively even filtrations $F_s^{\cdot}(\underline{\underline{g}})$ with respect to $\underline{\underline{g}}$ on $W_s(\underline{\underline{g}}^m)$ $(s \geq 0,\ m \geq 1)$ such that $F_0^{\cdot}(\underline{\underline{g}})$ on $W_0(\underline{\underline{g}}) = W(\underline{\underline{g}})$ is given by the canonical filtration

$$F^{2p}W(\underline{\underline{g}}) = S^p(\underline{\underline{g}}^*).W(\underline{\underline{g}}), \quad F^{2p-1} = F^{2p}$$

Let

$$F_0^{2p}(\underline{\underline{g}})W(\underline{\underline{g}}^m) = \mathrm{Id}\{(W^+(\underline{\underline{g}}^m)^{i(\underline{\underline{g}})}\}^p$$

(8.7) $$F_s^{\cdot}(\underline{\underline{g}})W_s(\underline{\underline{g}}^m) = \underset{\ell \geq 0}{\oplus} F_s^{\cdot}(\underline{\underline{g}})W_s^{\ell}(\underline{\underline{g}}^m)$$

$$F_s^{\cdot}(\underline{\underline{g}})W_s^{\ell}(\underline{\underline{g}}^m) = F_{s-1}^{\cdot}(\underline{\underline{g}})W_{s-1}\{(\underline{\underline{g}}^m)^{\ell+1}\}, \quad s \geq 1.$$

The odd filtrations are defined by $F_s^{2p-1} = F_s^{2p}$. The face and degeneracy operators of W_s are filtration-preserving. The filtration F_s is functorial for maps $W_s(\underline{\underline{g}}) \longrightarrow W_s(\underline{\underline{g}}')$ induced by Lie homomorphisms $\underline{\underline{g}}' \longrightarrow \underline{\underline{g}}$. One verifies that $F_s^{\cdot}\pi_s$ is an even, bihomogeneous and multiplicative filtration by $\underline{\underline{g}}$-DG-ideals.

The split exact sequence

$$0 \longrightarrow \underline{\underline{g}} \xrightarrow{\Delta} \underline{\underline{g}}^{\ell+1} \longrightarrow V_\ell \longrightarrow 0$$

defines the $\underline{\underline{g}}$-module V_ℓ, whose dual is given by $V_\ell^* = \ker \Delta^* = \{(\alpha_0,\ldots,\alpha_\ell) \mid \sum_{i=0}^{\ell} \alpha_i = 0\}$. The filtration $F_1^{2p}W^\ell(\underline{\underline{g}}) = F_0^{2p}W(\underline{\underline{g}}^{\ell+1})$ is then given by

$$(8.8) \qquad F_1^{2p}W_1^\ell(\underline{\underline{g}}) \cong \bigoplus_{|r|\geq p} \Lambda^{\cdot}\underline{\underline{g}}^* \otimes \{\Lambda^{\cdot}V_\ell^* \otimes S^{\cdot}(\underline{\underline{g}}^{*\ell+1})\}^{|r|}$$

where the weight $|r|$ is determined by weight $\Lambda^1 V_\ell^* =$ = weight $S^1(\underline{\underline{g}}^{*\ell+1}) = 1$. For the graded object we have therefore

$$G_1^2 W_1^\ell(\underline{\underline{g}}) = \Lambda\underline{\underline{g}}^* \otimes \{\Lambda V_\ell^* \otimes S(\underline{\underline{g}}^{*\ell+1})\}^{|p|}$$

For every subalgebra $\underline{\underline{h}} \subset \underline{\underline{g}}$ the filtrations $F_s^{\cdot}$ induce filtrations on the relative algebras

$$(8.9) \qquad W_s(\underline{\underline{g}},\underline{\underline{h}}) = \{W_s(\underline{\underline{g}})\}_{\underline{\underline{h}}}, \qquad s \geq 0.$$

It is immediate that the canonical projections $\rho_s : W_s \longrightarrow W_{s-1}$ are filtration-preserving. Define for $s \geq 0$

$$(8.10) \qquad W_s(\underline{\underline{g}},\underline{\underline{h}})_k = W_s(\underline{\underline{g}},\underline{\underline{h}})/F_s^{2(k+1)}W_s(\underline{\underline{g}},\underline{\underline{h}}), \qquad k \geq 0.$$

For $k = \infty$ we set $F^\infty = \bigcap_{p\geq 0} F^{2p} = 0$, so that

$$(8.11) \qquad W_s(\underline{\underline{g}},\underline{\underline{h}})_\infty = W_s(\underline{\underline{g}},\underline{\underline{h}}).$$

The main result concerning the relationship between the W_s is then as follows [KT 6,7,8,12].

8.12 THEOREM. Let $(\underline{\underline{g}},\underline{\underline{h}})$ be a pair of Lie algebras. The homomorphisms of spectral sequences induced by the filtration-preserving canonical projections $\rho_s : W_s(\underline{\underline{g}},\underline{\underline{h}}) \longrightarrow W_{s-1}(\underline{\underline{g}},\underline{\underline{h}})$ induce isomorphisms on the E_2-level and hence isomorphisms for every $0 \leq k \leq \infty$

$$H(\rho_s) : H(W_s(\underline{\underline{g}},\underline{\underline{h}})_k) \xrightarrow{\cong} H(W_{s-1}(\underline{\underline{g}},\underline{\underline{h}})_k), \qquad s > 0.$$

The E_2-term of the spectral sequence for $s = 0$ has been determined for reductive pairs in [KT 5] to be

$$E_2^{2p,q} = H^q(\underline{\underline{g}},H) \otimes I(\underline{\underline{g}})_k^{2p}.$$

It follows that there exists an even multiplicative spectral sequence

$$(8.13) \qquad E_2^{2p,q} = H^q(\underline{\underline{g}},\underline{\underline{h}}) \otimes I(\underline{\underline{g}})_k^{2p} \Longrightarrow H^{2p+q}(W_s(\underline{\underline{g}},\underline{\underline{h}})_k)$$

for $s \geq 0$, $0 \leq k \leq \infty$.

With the $\underline{\underline{g}}$-DG-map $\lambda : W(\underline{\underline{g}}) \longrightarrow W_1(\underline{\underline{g}})$ of chapter 5, theorem 5.35 we have then the following consequence.

8.14 COROLLARY. λ induces a mapping of complexes

$$\lambda : W(\underline{\underline{g}},\underline{\underline{h}})_k \longrightarrow W_1(\underline{\underline{g}},\underline{\underline{h}})_k$$

for any subalgebra $\underline{\underline{h}} \subset \underline{\underline{g}}$, $0 \leq k \leq \infty$ and we have $H(\lambda) = H(\rho_1)^{-1}$. Thus $H(\lambda)$ is multiplicative.

We give some explicit formulas for λ:

$$(8.15) \qquad \begin{aligned} \lambda(\alpha) &= (\alpha,0,\ldots) & \alpha &\in \Lambda^1(\underline{\underline{g}}^*) = W^{1,0}, \\ \lambda(\tilde{\alpha}) &= (\tilde{\alpha},\delta\alpha,0,\ldots) & \tilde{\alpha} &\in S^1(\underline{\underline{g}}^*) = W^{0,2}, \end{aligned}$$

where $\delta\alpha = 1 \otimes \alpha - \alpha \otimes 1$. For $\alpha \in \Lambda^1(\underline{\underline{g}}^*)^{\underline{\underline{g}}}$ and $\tilde{\alpha}$ the corresponding element in $S^1(\underline{\underline{g}}^*)^{\underline{\underline{g}}} = I(\underline{\underline{g}})^2$:

(8.16) $$\lambda(\alpha\tilde{\alpha}) = \alpha\tilde{\alpha}, -(\alpha\otimes 1)\delta\alpha, 0, \ldots).$$

Thus $\lambda(\alpha\tilde{\alpha}) = \lambda(\alpha).\lambda(\tilde{\alpha}) = \lambda(\tilde{\alpha}).\lambda(\alpha)$ in this case and hence also

(8.17) $$\lambda(\alpha\tilde{\alpha}^q) = \lambda(\alpha).\lambda(\tilde{\alpha})^q, \quad q \geq 1.$$

8.18 GENERALIZED WEIL HOMOMORPHISM. We consider a foliated G-bundle $P \longrightarrow M$, $\mathfrak{U} = (U_j)$ a sufficiently fine open covering of M and $\omega = (\omega_j)$ a family of connections ω_j in $P|U_j$ representing the flat partial connection in P. These data define then a homomorphism

(8.19) $$k_1(\omega) : W_1^{\cdot}(\underline{\underline{g}}) \longrightarrow \check{C}^{\cdot}(\mathfrak{U}, \pi_*\Omega_P^{\cdot})$$

as follows. For $\ell \geq 0$, let $\sigma = (i_0, \ldots, i_\ell)$ be an ℓ-simplex of the nerve $N(\mathfrak{U})$. Consider the compositions

$$\omega_{i_j} : \Lambda(\underline{\underline{g}}^*) \longrightarrow \Gamma\{U_{i_j}, \pi_*\Omega_P^{\cdot}\} \longrightarrow \Gamma(U_\sigma, \pi_*\Omega_P^{\cdot}) \quad \text{for} \quad j = 0, \ldots, \ell.$$

This defines

(8.20) $$k(\omega_\sigma) : W(\underline{\underline{g}}^{\ell+1}) \longrightarrow \Gamma(U_\sigma, \pi_*\Omega_P^{\cdot})$$

as the universal $\underline{\underline{g}}$-DG-algebra homomorphism extending

(8.21) $$\Lambda(\omega_\sigma) : \Lambda(\underline{\underline{g}}^{*\ell+1}) \longrightarrow \Gamma(U_\sigma, \pi_*\Omega_P^{\cdot})$$

given on the factor j by ω_{i_j}. We get therefore a homomorphism $k_1(\omega) : W_1^{\ell}(\underline{\underline{g}}) \longrightarrow \check{C}^{\ell}(\mathfrak{U}, \pi_*\Omega_P^{\cdot})$ by setting $k_1(\omega)_\sigma = k(\omega_\sigma)$. We observe that this homomorphism is the composition of the following two $\underline{\underline{g}}$-DG-homomorphisms. The unique map of semi-simplicial sets $N(\mathfrak{U}) \longrightarrow Pt$ first pulls the canonical system W on Pt back to a local system $\underset{\sim}{W}_{\mathfrak{U}}$ on $N(\mathfrak{U})$. Then the assignment $\sigma \longrightarrow k(\omega_\sigma)$ defines a map of local systems $\underset{\sim}{W}_{\mathfrak{U}} \longrightarrow \pi_*\Omega_P^{\cdot}$. The induced homomorphisms compose to $k_1(\omega)$. Thus (8.19) is a homomorphism of

$\underline{\underline{g}}$-DG-algebras, the generalized Weil-homomorphism of P. Together with (8.12) the following result is crucial for our construction.

8.22 PROPOSITION. $k_1(\omega)$ <u>is filtration-preserving in the sense that</u>

$$k_1(\omega) : F_1^{2p}W_1^{\cdot} \longrightarrow F^p\check{C}^{\cdot}(\mathfrak{U},\pi_*\Omega_P^{\cdot}),\ p \geq 0.$$

Proof. The filtration on the image complex is defined by

$$(8.23)\qquad F^p\check{C}^{\cdot}(\mathfrak{U},\pi_*\Omega_P^{\cdot}) = \check{C}^{\cdot}(\mathfrak{U},F^p\Omega_P^{\cdot}) = \check{C}^{\cdot}\{\mathfrak{U},(\Omega.\pi_*\Omega_P^{\cdot})^p\}.$$

Similarly $\check{C}^{\cdot}(\mathfrak{U},\Omega_M^{\cdot})$ is filtered by

$$(8.24)\qquad F^p\check{C}^{\cdot}(\mathfrak{U},\Omega_M^{\cdot}) = \check{C}^{\cdot}(\mathfrak{U},F^p\Omega_M^{\cdot}) = \check{C}^{\cdot}\{\mathfrak{U},(\Omega.\Omega_M)^p\}.$$

By the multiplicativity of the filtration and (8.8) it is sufficient to verify the relations

$$k(\omega_\sigma)\, u \in \Gamma(U_\sigma,F^1\pi_*\Omega^2) \quad \text{for} \quad u \in S^1(\underline{\underline{g}}^{*\ell+1})$$

and

$$k(\omega_\sigma)\, v \in \Gamma(U_\sigma,F^1\pi_*\Omega_P^1) \quad \text{for} \quad v \in \Lambda^1 V_\ell^*.$$

But $S^1(\underline{\underline{g}}^{*\ell+1})$ and $\Lambda^1 V_\ell^*$ are generated linearly by the elements $\tilde{\alpha}_j = (0,\dots,\overset{j}{\tilde{\alpha}},\dots,0)$, $\tilde{\alpha} \in S^1(\underline{\underline{g}}^*)$ and $\alpha_{jk} = (0,\dots,\overset{j}{-\alpha},\dots,\overset{k}{\alpha},\dots,0)$ for $\alpha \in \Lambda^1\underline{\underline{g}}^*$ and it is sufficient to check the relations on these elements. Now for $\sigma = (i_0,\dots,i_\ell)$ we get $k(\omega_\sigma)\tilde{\alpha}_j = K(\omega_{i_j})\tilde{\alpha} \in \Gamma(U_{i_j},F^1\pi_*\Omega_P^2)$ and $k(\omega_\sigma)\alpha_{jk} = (\omega_{i_k}-\omega_{i_j})\alpha \in \Gamma(U_{i_j,i_k},F^1\pi_*\Omega_P^1)$ since the local connections ω_i in $P|U_i$ are adapted to the given flat partial connection in P.

Let now q be the codimension of the foliation on M. It is then clear that the filtrations (8.23) (8.24) are zero for $p > q$. Thus $k_1(\omega)$ induces by (8.22) a homomorphism

$$k_1(\omega) : W_1^{\cdot}(\underline{\underline{g}})_q \longrightarrow \check{C}^{\cdot}(\mathcal{U},\pi_*\Omega_P^{\cdot}).$$

More generally for a closed subgroup $H \subset G$ with finitely many connected components and Lie algebra $\underline{\underline{h}} \subset \underline{\underline{g}}$ we have an induced map between the H-basic algebras of (8.19). If $\hat{\pi} : P/H \longrightarrow M$ denotes the projection induced from $\pi : P \longrightarrow M$, then $(\pi_*\Omega_P^{\cdot})_H = \hat{\pi}_*\Omega_{P/H}^{\cdot}$ and hence

$$k_1(\omega) : W_1(\underline{\underline{g}},H) \longrightarrow \check{C}(\mathcal{U},\hat{\pi}_*\Omega_{P/H})$$

Since this map is still filtration-preserving, and the filtration on the RHS is zero for degrees exceeding q, we get

$$(8.25) \qquad k_1(\omega) : W_1(\underline{\underline{g}},H)_q \longrightarrow \check{C}(\mathcal{U},\hat{\pi}_*\Omega_{P/H}).$$

To define invariants in the base manifold M, we need as before an H-reduction of P given by a section $s : M \longrightarrow P/H$ of $\hat{\pi} : P/H \longrightarrow M$ as the pull-back $P' = s^*P$. It defines a homomorphism $s^* : \check{C}^{\cdot}(\mathcal{U},\hat{\pi}_*\Omega_{P/H}^{\cdot}) \longrightarrow \check{C}^{\cdot}(\mathcal{U},\Omega_M^{\cdot})$. Note that the cohomology of this target complex maps canonically into the De Rham cohomology of M (viewed as hypercohomology): $j : H^{\cdot}(\check{C}(\mathcal{U},\Omega_M^{\cdot})) \longrightarrow \mathbf{H}^{\cdot}(M;\Omega_M^{\cdot}) = H_{DR}^{\cdot}(M)$. Thus we can finally write down the definition of the generalized characteristic homomorphism

$$\Delta_* : H^{\cdot}(W(\underline{\underline{g}},H)_q) \longrightarrow H_{DR}^{\cdot}(M)$$

of a foliated bundle P equipped with an H-reduction s^*P as

$$(8.26) \qquad \Delta_* = js^*(k_1)_*\lambda_* \, .$$

This extends theorem 4.43 to holomorphic and algebraic bundles [KT 6,7,8,12]. The only additional assumption needed in the present situation is that the closed subgroup $H \subset G$ has finitely many connected components.

Observe that with the definition (5.51) of $\lambda_{\underset{\sim}{E}} : W \longrightarrow \check{C}(\mathcal{U},\underset{\sim}{E})$ and $\lambda : W \longrightarrow W_1$ as in theorem 5.35 we have now the relation

$$k_1(\omega) \circ \lambda = \lambda_{\underset{\sim}{E}} ,$$

where $\underset{\sim}{E}$ is the local system of $\underline{\underline{g}}$-DG-algebras given by $\Gamma(-,\pi_*\Omega^{\cdot}_P)$.

Note that one has in fact also in this case more generally a homomorphism of filtered $\underline{\underline{g}}$-DG-algebras $\Delta(\omega) : W_1(\underline{\underline{g}},H)_q \longrightarrow \check{C}(\mathcal{U},\Omega^{\cdot}_M)$. This map induces a homomorphism of the associated multiplicative spectral sequences. The spectral sequence converging to $H(W_1(\underline{\underline{g}},h)_q)$ is given in (8.13). The spectral sequence of $H^{\cdot}_{DR}(M)$ associated to the filtration (8.24) is of the general form [KT 4,7]

$$E_1^{s,t}(Q^*) = \mathrm{Ext}^t_{\underline{U}(\underline{L})}(M;\underline{O}_M,\Lambda^s\underline{Q}^*) \Longrightarrow H^{s+t}_{DR}(M) \tag{8.27}$$

where $\underline{L} = (\Omega^1_M/\underline{Q}^*)^* \subset \underline{T}_M$ is the annihilator sheaf of $\underline{Q}^*$ in $\underline{T}_M$, and $\underline{U}(\underline{L})$ is the universal envelope of the sheaf of Lie algebras $\underline{L}$. This spectral sequence is the Leray spectral sequence for De Rham cohomology in the case where the foliation on M is defined by a global submersion, and it may therefore be considered as a proper generalization of the Leray spectral sequence to foliations. It takes a more familiar form in other cases as well. The filtration preserving homomorphism $\Delta(\omega)$ induces then a multiplicative morphism of the spectral sequences as indicated in the following diagram

$$\begin{array}{ccc} H^{2s+t}(W_1(\underline{\underline{g}},H)_q) & \xrightarrow{\Delta_*} & H^{2s+t}_{DR}(M) \\ \Uparrow & & \Uparrow \\ E_2^{2s,t} = H^t(\underline{\underline{g}},H) \otimes I(G)^{2s}_q & \xrightarrow{\Delta_1^{s,t}} & E_1^{s,s+t}(Q^*) \end{array} \tag{8.28}$$

The homomorphism Δ_1 is called the first derived characteristic homomorphism of the foliated bundle in question. On $I(G)_q = E_2^0$, the map Δ_1 gives a generalization of the characteristic classes of a complex-analytic principal bundle in Hodge cohomology defined by Atiyah in [AT], while the invariants defined by Δ_1 on $E_2^{0,\cdot} = H^\cdot(\underline{\underline{g}},H)$ are invariants associated to the partial flat structure in P given by the foliation. All this has been explained in detail in [KT 7].

As an example consider a G-principal bundle $P \longrightarrow M$ with its unique (up to homotopy) K-reduction for a maximal compact subgroup $K \subset G$. Thus we consider the point foliation on M and P. The spectral sequence (8.13) is now of the form $(m = 2p+q)$:

$$(8.29) \qquad E_2^{2p,q} = H^q(\underline{\underline{g}},K) \otimes I(g)^{2p} \Longrightarrow H^m(W_1(\underline{\underline{g}},K)) \cong I(K)^m$$

by (8.12) and [KT 5, Prop. 1, (iii)]. The composition of the edge homomorphism $I(G)^{2p} \longrightarrow H^{2p}(W_1(\underline{\underline{g}},K))$ with Δ_* is precisely the usual Chern-Weil homomorphism of the principal bundle P constructed via Čech-cohomology, i.e. using local connection data only. For the universal bundle $E_G \longrightarrow B_G$ the homomorphism Δ_* turns out to be an isomorphism since it is contained in the diagram

$$(8.30) \qquad \begin{array}{ccc} H^\cdot(W_1(\underline{\underline{g}},K)) & \xrightarrow{\Delta_*} & H^\cdot(B_G,\mathbb{R}) \\ \downarrow\cong & & \downarrow\cong \\ I(K) & \xrightarrow{\cong} & H^\cdot(B_K,\mathbb{R}). \end{array}$$

Hence we obtain a spectral sequence

$$(8.31) \quad E_2^{2p,q} = H^q(\underline{\underline{g}},K) \otimes I(G)^{2p} \Longrightarrow H^m(W_1(\underline{\underline{g}},K)) \cong H^m(B_G,\mathbb{R})$$

with edge-homomorphism given by the universal Chern-Weil homomorphism $I(G) \longrightarrow H(B_G, \mathbb{R})$. This spectral sequence and the semisimplicial realization of $H^\cdot(B_G, \mathbb{R})$ coincide with the result in [B 5], [SH 1] via the van Est Theorem [E].

8.32 APPLICATION TO HOLOMORPHIC BUNDLES. As another application we consider the case of a foliated vectorbundle in the complex analytic category. Let M be a complex manifold with a complex-analytic foliation, and let E^r be a holomorphic vector-bundle of rank r such that the holomorphic frame bundle $F(E)$ is foliated. This means that E is equipped with a holomorphic partial flat connection along the Lie algebra subsheaf $\underline{L} = (\Omega^1_M/\underline{Q}^*)^* \subset \underline{T}_M$ defined by Q^*. If the holomorphic line bundle $\wedge^r E$ admits a non-zero holomorphic section, i.e. a holomorphic section $s : M \longrightarrow F(E)/SL(r,\mathbb{C}) = F(\wedge^r E)$, we obtain by our general procedure a characteristic homomorphism

(8.33) $$\Delta_* : H^\cdot(W(\underline{\underline{gl}}(r,\mathbb{C}), \underline{\underline{sl}}(r,\mathbb{C}))_q) \longrightarrow H^\cdot_{DR}(M)$$

where $q = \text{rank}\ \underline{Q}^*$ over $\underline{O}$ is the complex codimension. By the computations in chapter 5 or [KT 5,11] we have then

(8.34) $$H^i(W(\underline{\underline{gl}}(r,\mathbb{C}),\underline{\underline{sl}}(r,\mathbb{C}))_q) \cong \begin{cases} \mathbb{C}[c_2, \dots ,c_r]^i_q, & i \leq 2q, \\ \mathbb{C}\alpha \otimes \mathbb{C}[c_1, \dots ,c_r]^{2q}, & i = 2q+1, \\ 0 & , \quad i > 2q+1, \end{cases}$$

where $\tilde{c}_j(A) = tr(\wedge^j(\frac{i}{2\pi} A))$ represents the Chern polynomial in $I(GL(r,\mathbb{C}))$ and $\alpha = \frac{i}{2\pi}\ tr \in (\wedge^1 \underline{\underline{gl}}^*)^{\underline{\underline{gl}}} \subset W^{1,0}$, $d\alpha = \tilde{c}_1$. The $\Delta_*(\tilde{c}_j) \in H^j_{DR}(M)$, $j \geq 2$, are the Chern-classes of E and for $c_\lambda = \tilde{c}_1^{\lambda^1} \dots \tilde{c}_r^{\lambda^1}$, $\sum_{j=1}^{r} j.\lambda_j = q$ the classes $\Delta_*(\alpha \otimes c_\lambda)$ are secondary invariants of the foliated holomorphic bundle E. The Čech-cocycle in $C = \check{C}^\cdot(\mathcal{U},\Omega^\cdot_M)$ representing the class $\Delta_*(\alpha \otimes \tilde{c}_1^q)$ can be computed as follows. Let $\mathcal{U} = (U_j)$ be a Γ-acyclic open covering of M and choose local holomorphic frames $\bar{s}_i = (\bar{s}_i^1, \dots ,\bar{s}_i^r)$ of $E|U_i$. By assumption we may adjust the frames $\bar{s}_i$ so that the transition functions $\bar{g}_{ij} : U_{ij} \longrightarrow GL(r,\mathbb{C})$ given by $\bar{s}_j = \bar{s}_i \cdot \bar{g}_{ij}$ satisfy $\det(\bar{g}_{ij}) = 1$. The local sections $s_i = \overline{s}_i^1 \wedge \ldots \wedge s_i^r$ define then a global trivialization of $\wedge_E^r$. A family of local holomorphic connections (ω_i) representing the given foliation on $F(E)$ is then determined by the local $\underline{\underline{gl}}$-valued connection forms $\overline{\theta}_i = s_i^* \omega_i$ on U_i. For $\theta_i = tr(\overline{\theta}_i)$ we have the relations $d\theta_i \in \Gamma(U_i, \underline{Q}^* \Omega_M^1)$ and $\theta_{ij} = \theta_j - \theta_i \in \Gamma(U_{ij}, Q^*)$.

For the chains $u = \frac{i}{2\pi}(\theta_j) \in \check{C}^{0,1}$ and $v = \frac{i}{2\pi}(d\theta_j, \theta_{jk}) \in \check{C}^{0,2} \otimes \check{C}^{1,1}$ satisfying $du = v$, we have by (8.15) (8.16) $\Delta(\omega)\alpha = u$, $\Delta(\omega)\tilde{c}_1 = v$ and hence by (8.17)

$$(8.35) \qquad \Delta(\omega)(\alpha \otimes \tilde{c}_1^q) = u.v^q.$$

Explicitly for $q = 1$:

$$(8.36) \qquad \Delta(\omega)(\alpha \otimes \tilde{c}_1) = -(2\pi)^{-2}(\theta_j \wedge d\theta_j, -\theta_j \wedge \theta_{jk}) \in F^1C^3.$$

We can actually do better than that. It is well-known that the class $\tilde{c}_1(E) \in H^2(M, \mathbb{Z})$ can be realized as coboundary of the exact sheaf sequence

$$0 \longrightarrow \mathbb{Z} \longrightarrow \underline{O}_M \xrightarrow{\exp} \underline{O}_M^* \longrightarrow 1$$

where $\underline{O}^*$ is the sheaf of units in the structure sheaf $\underline{O}_M$ and $\exp = e^{2\pi i -}$. On a simple covering $\mathfrak{U}$ the class $\tilde{c}_1(M) \in H^2(M, Z)$ is then represented by the 2-cycle $c = \frac{i}{2\pi}(-\delta \log g) \in \check{C}^2(\mathfrak{U}, \mathbb{Z})$ where $g_{ik} = (\det \overline{g}_{ik}) \in \check{C}^1(\mathfrak{U}, \underline{O}^*)$. Hence if $\tilde{c}_1(E) = 0$ we may adjust the local sections $\overline{s}_i$ of $F(E)$ such that $\delta(\log g) = 0$ for a convenient choice of $\log g_{ij}$. If we define $u = \frac{i}{2\pi}(\theta_j, \log g_{jk}) \in \check{C}^1$ and $v = \frac{i}{2\pi}(d\theta_j, \theta_{jk}) \in F^1\check{C}^2$, where now $\theta_{jk} = \theta_k - \theta_j - d\log g_{jk}$, we have the general relation $d_{\check{C}}(u) = v - c$ and hence by our assumption $d_{\check{C}}(u) = v$. As $v^{q+1} \in F^{q+1}\check{C}$ is necessarily zero, it follows that $u.v^q$ is closed and defines a cohomology class in $H_{DR}^{2q+1}(M) \cong H^{2q+1}(M, \mathbb{C})$. For $q = 1$ the cocycle $u \cdot v$ is explicitly given by

(8.37) $u \cdot v = \frac{-1}{4\pi^2}(\theta_j \wedge d\theta_j, \ -\theta_j \wedge \theta_{jk} + \log g_{jk} \cdot d\theta_k, \ \log g_{jk} \cdot \theta_{kl}) \in \check{C}^3$

Hence $u \cdot v$ is composed of chains of bidegree (0,3), (1,2) and (2,1) respectively.

More generally we have for $q > 0$:

8.38 PROPOSITION. <u>If</u> $\tilde{c}_1(E) \in H^2(M,\mathbb{Z})$ <u>is zero, the characteristic homomorphism</u> Δ_* <u>in</u> (8.33) (8.34) <u>is still defined. The classes</u> $\Delta_*(\alpha \otimes c_\lambda)$ <u>satisfying</u> $\lambda_1 > 0$ <u>are independent of all choices made and therefore define invariants of the foliated holomorphic bundle</u> E.

8.39 COROLLARY. <u>If</u> $\Lambda^r E$ <u>admits a global non-zero holomorphic section</u> s, <u>then the characteristic homomorphism</u> Δ_* <u>is independent of</u> s <u>on the classes</u> $\alpha \otimes c_\lambda$, $\lambda_1 > 0$.

If we take for E the normal bundle Q* equipped with

REFERENCES

[A] V. I. Arnold: Characteristic class entering in quantization conditions, Funct. Anal. and its Appl. 1(1967), 1-13.

[AT] M. F. Atiyah: Complex analytic connections in fibre bundles, Trans. Amer. Math. Soc. 85(1957), 181-207.

[B 1] R. Bott: On a topological obstruction to integrability, Proc. Symp. in Pure Math., Amer. Math. Soc. 16(1970), 127-131.

[B 2] ————: On topological obstructions to integrability, Proc. Internat. Congress Math. Nice (1970), Vol. 1, Gauthier-Villars, Paris (1971), 27-36.

[B 3] ————: Lectures on characteristic classes and foliations, Lectures on Algebraic and Differential Topology, Springer Lecture Notes, 279(1972), 1-94.

[B 4] ————: The Lefschetz formula and exotic characteristic classes, Symp. Math. Vol. X, Rome (1972), 95-105.

[B 5] ————: On the Chern-Weil homomorphism and the continuous cohomology of Lie groups, Advances in Math 11(1973), 289-303.

[BH] R. Bott and A. Haefliger: On characteristic classes of Γ-foliations, Bull. Amer. Math. Soc. 78(1972), 1039-1044.

[BO 1] A. Borel: Sur la cohomologie des espaces fibrés principaux et des espaces homogènes de groupes de Lie compacts, Ann. of Math. 57(1953), 115-207.

[BO 2] ————: Compact Clifford-Klein forms of symmetric spaces, Topology 2(1963), 111-122.

[BR 1] I. N. Bernstein and B. I. Rosenfeld: Characteristic classes of foliations, Funct. Anal. and its Appl. 6(1972), 68-69.

[BR 2] ————————————: Homogeneous spaces of infinite-dimensional Lie algebras and characteristic classes of foliations, Russ. Math. Surveys 28(1973), 107-142.

[C 1] S. S. Chern: Characteristic classes of Hermitian manifolds, Annals of Math. 47(1946), 85-121.

[C 2] ————: Topics in differential geometry, Inst. for Adv. Study, Princeton (1951).

[C 3] ————: Geometry of characteristic classes, Canad. Summer School, Halifax (1971).

[CA] H. Cartan: Cohomologie réelle d'un espace fibré principal differentiable, Séminaire Cartan, exposés 19 et 20 (1949/50).

[CL 1] Conlon, L: Transversally parallelizable foliations of codimension two, Trans. Amer. Math. Soc. 194(1974), 79-102.

[CL 2] ————: Foliations and locally free transformation groups of codimension two, Mich. Math. J. 21(1974), 87-96.

[CN] B. Cenkl: Secondary characteristic classes, Journées exotiques, Lille (1973).

[CS 1] S. S. Chern and J. Simons: Some cohomology classes in principal fibre bundles and their applications to Riemannian geometry, Proc. Nat. Acad. Sc. USA 68(1971), 791-794.

[CS 2] ————————: Characteristic forms and geometric invariants, Annals of Math. 99(1974), 48-69.

[E] W. T. van Est: Une application d'une methode de Cartan-Leray, Indag. Math. 18(1955), 542-544.

[ER] Ch. Ehresmann and G. Reeb: Sur les champs d'éléments de contact de dimension p complètement intégrables, C. R. Acad. Sc. Paris 218(1944), 955-957.

[GB 1] C. Godbillon: Problèmes d'existence et d'homotopie dans les feuilletages, Séminaire Bourbaki, 23e année 1970/71, exposé 390.

[GB 2] ————: Cohomologies d'algèbres de Lie de champs de vecteurs, Seminaire Bourbaki, 25e année (1971/72), no. 421.

[GF 1] I. M. Gelfand and D. B. Fuks: The cohomology of the Lie algebra of tangent vectorfields on a smooth manifold. J.

[H 1] A. Haefliger: Structures feuilletées et cohomologie à valeur dans un faisceau de groupoïdes, Comm. Math. Helv. 32(1958), 248-329.

[H 2] ————: Variétés feuilletées, Ann. Scuola Norm. Sup. Pisa 16(1962), 367-397.

[H 3] ————: Homotopy and integrability, Manifolds-Amsterdam 1970, Springer Lect. Notes in Math., no. 197, 133-166.

[H 4] ————: Feuilletages sur les variétés ouvertes, Topology 9(1970), 183-194.

[H 5] ————: Sur les classes caractéristiques des feuilletages, Séminaire Bourbaki, 24 e année 1971/72, n° 412.

[H 6] ————: Sur la cohomologie de Gelfand-Fuks, Journées Differentielles, Dijon (1974).

[HG] S. Helgason: Differential geometry and symmetric spaces, Academic Press, New York (1962).

[HL] J. Heitsch and H. B. Lawson: Transgressions, Chern-Simons invariants and the classical groups, J. of Diff. Geom. 9(1974), 423-434.

[HO 1] H. Hopf: Über die Curvatura Integra geschlossener Hyperflächen, Math. Ann. 95(1925/26), 340-367.

[HO 2] ————: Vektorfelder in n-dimensionalen Mannigfaltigkeiten, Math. Ann. 96(1926/27), 225-250.

[HR] F. Hirzebruch: Automorphe Formen und der Satz von Riemann-Roch, Symp. Intern. de Topologia Algebraica, Mexico (1958), 129-144.

[HT] J. Heitsch: Deformations of secondary characteristic classes, Topology 12(1973), 381-388.

[K 1] J. L. Koszul: Homologie et cohomologie des algebres de Lie, Bull. Soc. Math. France 78(1950), 65-127.

[K 2] ————: Sur un type d'algèbres différentielles en rapport avec la transgression, Coll. de Topologie, Bruxelles (1950), p. 73-81.

[K 3] ————: Multiplicateurs et classes caractéristiques, Trans. Amer. Math. Soc. 89(1958), 256-266.

[K 4] ————: Espaces fibrés et pré-associés, Nagoya Math. J. 15(1959), 155-169.

[K 5] ————: Déformations et connexions localement plates, Ann. Inst. Fourier, Grenoble 18(1968), 103-114.

[K 6] ————: Connexions L-equivalentes et formule de Chern-Simons, d'après Kobayashi-Ochiai, Journées Exotiques, Lille (1973).

[KN 1] S. Kobayashi and K. Nomizu: Foundations of differential geometry, vol. I (1963), Interscience, New York.

[KN 2] ——————————: Foundations of differential geometry, vol. II (1969), Interscience, New York.

[KO] S. Kobayashi and T. Ochiai: G-structures of order two and transgression operators, J. of Diff. Geom. 6(1971), 213-230.

[KT 1] F. Kamber and Ph. Tondeur: Flat manifolds, Lecture Notes in Mathematics 67(1968), Springer, Berlin.

[KT 2] ——————————: Invariant differential operators and cohomology of Lie algebra sheaves, Differentialgeometrie im Grossen, Juli 1969, Berichte aus dem Math. Forschungsinstitut Oberwolfach, Heft 4, Mannheim (1971), 177-230.

[KT 3] ——————————: Invariant differential operators and cohomology of Lie algebra sheaves, Memoirs of the American Math. Soc., Number 113 (1971), 124 pp.

[KT 4] ——————————: Characteristic classes of modules over a sheaf of Lie algebras, Notices Amer. Math. Soc. 19, A-401 (1972).

[KT 5] ——————————: Cohomologie des algèbres de Weil relatives tronquées, C. R. Acad. Sc. Paris, t. 276 (1973), 459-462.

[KT 6] ——————————: Algèbres de Weil semi-simpliciales, C. R. Acad. Sc. Paris, t. 276(1973), 1177-1179; Homomorphisme caractéristique d'un fibré principal feuilleté, ibid. 1407-1410; Classes caractéristiques dérivées d'un fibré principal feuilleté, ibid,. 1449-1452.

[KT 7] ——————————: Characteristic invariants of foliated bundles, Manuscripta Mathematica 11(1974), 51-89.

[KT 8] ——————————: Semi-simplicial Weil algebras and characteristic classes for foliated bundles in Čech cohomology, Proc. Symposia Pure Math. Vol. 27, 283-294(1975).

[KT 9] ——————————: Classes caractéristiques generalisées des fibrés feuilletés localement homogènes, C. R. Acad. Sc. Paris t. 279(1974), 847-850; Quelques classes caractéristiques generalisées nontriviales de fibrés feuilletés, ibid., 921-924.

[KT 10] ——————————: Non-trivial characteristic invariants of homogeneous foliated bundles, to appear in Ann.Ec.Norm.Sup.

[KT 11] ——————————: Cohomology of g-DG-algebras, to appear.

[KT 12] ——————————: Semi-simplicial Weil algebras and characteristic classes, to appear.

[L] H. B. Lawson: Foliations, Bull. Amer. Math. Soc. 80(1974), 369-418.

[LK] M. V. Losik: Cohomologies of the Lie algebra of vector fields with coefficients in a trivial unitary representation, Functional Analysis 6(1972), 24-36.

[LN 1] D. Lehmann: J-homotopie dans les espaces de connexions et classes exotiques de Chern-Simons, C. R. Acad. Sc. Paris t. 275(1972), 835-838.

[LN 2] ————: Classes caractéristiques exotiques et J-connexité des espaces de connexions, Journées exotiques, Lille (1973).

[LP] C. Lazarov and J. Pasternack: Secondary characteristic classes for Riemannian foliations, to appear.

[MO 1] P. Molino: Classes d'Atiyah d'un feuilletage et connexions transverses projetables, C. R. Ac. Sc. Paris t. 272(1971), 779-781.

[MO 2] ————: Classes caractéristiques et obstructions d'Atiyah pour les fibrés principaux feuilletés, C. R. Ac. Sc. Paris t. 272 (1971), 1376-1378.

[MO 3] ————: Feuilletages et classes caractéristiques, Symp. Math. Vol. X, Rome (1972), 199-209.

[MO 4] ————: Propriétés cohomologiques et propriétés topologiques des feuilletages à connexion transverse projetable, Topology 12(1973), 317-325.

[MO 5] ————: La classe d'Atiyah d'un feuilletage comme cocycle de déformation infinitésimale, C. R. Ac. Sc. Paris t. 278 (1974), 719-721.

[MR] J. Martinet: Classes caractéristiques des systèmes de Pfaff, to appear.

[MS] V. P. Maslov: Théorie des perturbations et méthodes asymptotiques, Gauthier-Villars, Paris (1972).

[MT] Y. Matsushima: On Betti numbers of compact locally symmetric Riemannian manifolds, Osaka Math. J. 14(1962), 1-20.

[P 1] J. Pasternack: Topological obstructions to integrability and Riemannian geometry of foliations, Thesis, Princeton (1970).

[P 2] ————: Foliations and compact Lie group actions, Comm. Math. Helv. 46(1971), 467-477.

[R] G. Reeb: Sur certaines propriétés topologiques des variétés feuilletées, Actual. Sci. Ind. No. 1183, Hermann, Paris (1952).

[RE 1] B. Reinhart: Foliated manifolds with bundle-like metrics, Annals of Math. 69(1959), 119-132.

[RE 2] ————: Algebraic invariants of foliations, Symposium on differential equations and dynamical systems Warwick, Springer Lecture Notes 206, 1971, 119-120.

[RE 3] B. Reinhart: Holonomy invariants for framed foliations, Differential Geometry Colloquium, Santiago de Compostela 1972, Springer Lecture Notes 392, 47-52.

[RW] B. Reinhart and J. Wood: A metric formula for the Godbillon-Vey invariant for foliations, Proc. Amer. Math. Soc. 38(1973), 427-430.

[SH 1] H. Shulman: Characteristic classes of foliations, Ph.D. Thesis, University of California Berkeley (1972).

[SH 2] ————: Secondary obstructions to foliations, Topology 13(1974), 177-183.

[ST] H. Shulman and D. Tischler: Leaf invariants of foliations and the van Est isomorphism, to appear.

[T] S. L. Tan: Nullity and generalized characteristic classes of differentiable manifolds, Thesis, University of Illinois at Urbana-Champaign (1974).

[TH 1] W. Thurston: Noncobordant foliations of S^3, Bull. Amer. Math. Soc. 78(1972), 511-514.

[TH 2] ————: The theory of foliations of codimension greater than one, Comm. Math. Helv. 49(1974), 214-231.

[TH 3] ————: The existence of foliations, Notices Amer. Math. Soc. 21(1974), A-407.

[TS] D. Tischler: On fibering certain foliated manifolds over S^1, Topology 9(1970), 153-154.

[V 1] J. Vey: Sur la cohomologie de l'algèbre des champs de vecteurs sur une variété, C. R. Acad. Sc. Paris 273(1971), 850-852.

[V 2] ————: Quelques constructions relatives aux Γ-structures, C. R. Acad. Sc. Paris 276(1973), 1151-1153.

[VZ 1] I. Vaisman: Sur la cohomologie des variétés Riemanniennes feuilletées, C. R. Ac. Sc. Paris t. 268(1969), 720-723.

[VZ 2] ————: Sur une classe de complexes de cochaînes, Math. Ann. 194(1971), 35-42.

INDEX OF SYMBOLS

Vol. 342: Algebraic K-Theory II, "Classical" Algebraic K-Theory, and Connections with Arithmetic. Edited by H. Bass. XV, 527 pages. 1973.

Vol. 343: Algebraic K-Theory III, Hermitian K-Theory and Geometric Applications. Edited by H. Bass. XV, 572 pages. 1973.

Vol. 344: A. S. Troelstra (Editor), Metamathematical Investigation of Intuitionistic Arithmetic and Analysis. XVII, 485 pages. 1973.

Vol. 345: Proceedings of a Conference on Operator Theory. Edited by P. A. Fillmore. VI, 228 pages. 1973.

Vol. 346: Fučík et al., Spectral Analysis of Nonlinear Operators. II, 287 pages. 1973.

Vol. 347: J. M. Boardman and R. M. Vogt, Homotopy Invariant Algebraic Structures on Topological Spaces. X, 257 pages. 1973.

Vol. 348: A. M. Mathai and R. K. Saxena, Generalized Hypergeometric Functions with Applications in Statistics and Physical Sciences. VII, 314 pages. 1973.

Vol. 349: Modular Functions of One Variable II. Edited by W. Kuyk and P. Deligne. V, 598 pages. 1973.

Vol. 350: Modular Functions of One Variable III. Edited by W. Kuyk and J.-P. Serre. V, 350 pages. 1973.

Vol. 351: H. Tachikawa, Quasi-Frobenius Rings and Generalizations. XI, 172 pages. 1973.

Vol. 352: J. D. Fay, Theta Functions on Riemann Surfaces. V, 137 pages. 1973.

Vol. 353: Proceedings of the Conference on Orders, Group Rings and Related Topics. Organized by J. S. Hsia, M. L. Madan and T. G. Ralley. X, 224 pages. 1973.

Vol. 354: K. J. Devlin, Aspects of Constructibility. XII, 240 pages. 1973.

Vol. 355: M. Sion, A Theory of Semigroup Valued Measures. V, 140 pages. 1973.

Vol. 356: W. L. J. van der Kallen, Infinitesimally Central Extensions of Chevalley Groups. VII, 147 pages. 1973.

Vol. 357: W. Borho, P. Gabriel und R. Rentschler, Primideale in Einhüllenden auflösbarer Lie-Algebren. V, 182 Seiten. 1973.

Vol. 358: F. L. Williams, Tensor Products of Principal Series Representations. VI, 132 pages. 1973.

Vol. 359: U. Stammbach, Homology in Group Theory. VIII, 183 pages. 1973.

Vol. 360: W. J. Padgett and R. L. Taylor, Laws of Large Numbers for Normed Linear Spaces and Certain Fréchet Spaces. VI, 111 pages. 1973.

Vol. 361: J. W. Schutz, Foundations of Special Relativity: Kinematic Axioms for Minkowski Space-Time. XX, 314 pages. 1973.

Vol. 362: Proceedings of the Conference on Numerical Solution of Ordinary Differential Equations. Edited by D.G. Bettis. VIII, 490 pages. 1974.

Vol. 363: Conference on the Numerical Solution of Differential Equations. Edited by G. A. Watson. IX, 221 pages. 1974.

Vol. 364: Proceedings on Infinite Dimensional Holomorphy. Edited by T. L. Hayden and T. J. Suffridge. VII, 212 pages. 1974.

Vol. 365: R. P. Gilbert, Constructive Methods for Elliptic Equations. VII, 397 pages. 1974.

Vol. 366: R. Steinberg, Conjugacy Classes in Algebraic Groups (Notes by V. V. Deodhar). VI, 159 pages. 1974.

Vol. 367: K. Langmann und W. Lütkebohmert, Cousinverteilungen und Fortsetzungssätze. VI, 151 Seiten. 1974.

Vol. 368: R. J. Milgram, Unstable Homotopy from the Stable Point of View. V, 109 pages. 1974.

Vol. 369: Victoria Symposium on Nonstandard Analysis. Edited by A. Hurd and P. Loeb. XVIII, 339 pages. 1974.

Vol. 370: B. Mazur and W. Messing, Universal Extensions and One Dimensional Crystalline Cohomology. VII, 134 pages. 1974.

Vol. 371: V. Poenaru, Analyse Différentielle. V, 228 pages. 1974.

Vol. 372: Proceedings of the Second International Conference on the Theory of Groups 1973. Edited by M. F. Newman. VII, 740 pages. 1974.

Vol. 373: A. E. R. Woodcock and T. Poston, A Geometrical Study of the Elementary Catastrophes. V, 257 pages. 1974.

Vol. 374: S. Yamamuro, Differential Calculus in Topological Linear Spaces. IV, 179 pages. 1974.

Vol. 375: Topology Conference. Edited by R. F. Dickman Jr. and P. Fletcher. X, 283 pages 1974.

Vol. 376: I. J. Good and D. B. Osteyee, Information, Weight of Evidence. The Singularity between Probability Measures and Signal Detection. XI, 156 pages. 1974.

Vol. 377: A. M. Fink, Almost Periodic Differential Equations. VIII, 336 pages. 1974.

Vol. 378: TOPO 72 - General Topology and its Applications. Proceedings 1972. Edited by R. A. Alò, R. W. Heath and J. Nagata. XIV, 651 pages. 1974.

Vol. 379: A. Badrikian et S. Chevet, Mesures Cylindriques, Espaces de Wiener et Fonctions Aléatoires Gaussiennes. X, 383 pages. 1974.

Vol. 380: M. Petrich, Rings and Semigroups. VIII, 182 pages. 1974.

Vol. 381: Séminaire de Probabilités VIII. Edité par P. A. Meyer. IX, 354 pages. 1974.

Vol. 382: J. H. van Lint, Combinatorial Theory Seminar Eindhoven University of Technology. VI, 131 pages. 1974.

Vol. 383: Séminaire Bourbaki - vol. 1972/73. Exposés 418-435. IV, 334 pages. 1974.

Vol. 384: Functional Analysis and Applications, Proceedings 1972. Edited by L. Nachbin. V, 270 pages. 1974.

Vol. 385: J. Douglas Jr. and T. Dupont, Collocation Methods for Parabolic Equations in a Single Space Variable (Based on C^1-Piecewise-Polynomial Spaces). V, 147 pages. 1974.

Vol. 386: J. Tits, Buildings of Spherical Type and Finite BN-Pairs. X, 299 pages. 1974.

Vol. 387: C. P. Bruter, Eléments de la Théorie des Matroïdes. V, 138 pages. 1974.

Vol. 388: R. L. Lipsman, Group Representations. X, 166 pages. 1974.

Vol. 389: M.-A. Knus et M. Ojanguren, Théorie de la Descente et Algèbres d' Azumaya. IV, 163 pages. 1974.

Vol. 390: P. A. Meyer, P. Priouret et F. Spitzer, Ecole d'Eté de Probabilités de Saint-Flour III - 1973. Edité par A. Badrikian et P.-L. Hennequin. VIII, 189 pages. 1974.

Vol. 391: J. W. Gray, Formal Category Theory: Adjointness for 2-Categories. XII, 282 pages. 1974.

Vol. 392: Géométrie Différentielle, Colloque, Santiago de Compostela, Espagne 1972. Edité par E. Vidal. VI, 225 pages. 1974.

Vol. 393: G. Wassermann, Stability of Unfoldings. IX, 164 pages. 1974.

Vol. 394: W. M. Patterson, 3rd, Iterative Methods for the Solution of a Linear Operator Equation in Hilbert Space - A Survey. III, 183 pages. 1974.

Vol. 395: Numerische Behandlung nichtlinearer Integrodifferential- und Differentialgleichungen. Tagung 1973. Herausgegeben von R. Ansorge und W. Törnig. VII, 313 Seiten. 1974.

Vol. 396: K. H. Hofmann, M. Mislove and A. Stralka, The Pontryagin Duality of Compact O-Dimensional Semilattices and its Applications. XVI, 122 pages. 1974.

Vol. 397: T. Yamada, The Schur Subgroup of the Brauer Group. V, 159 pages. 1974.

Vol. 398: Théories de l'Information, Actes des Rencontres de Marseille-Luminy, 1973. Edité par J. Kampé de Fériet et C.-F. Picard. XII, 201 pages. 1974.

Vol. 399: Functional Analysis and its Applications. Proceedings 1973. Edited by H. G. Garnir, K. R. Unni and J. H. Williamson. II, 584 pages. 1974.

Vol. 400: A Crash Course on Kleinian Groups. Proceedings 1974. Edited by L. Bers and I. Kra. VII, 130 pages. 1974.

Vol. 401: M. F. Atiyah, Elliptic Operators and Compact Groups. V, 93 pages. 1974.

Vol. 402: M. Waldschmidt, Nombres Transcendants. VIII, 277 pages. 1974.

Vol. 403: Combinatorial Mathematics. Proceedings 1972. Edited by D. A. Holton. VIII, 148 pages. 1974.

Vol. 404: Théorie du Potentiel et Analyse Harmonique. Edité par J. Faraut. V, 245 pages. 1974.

Vol. 405: K. J. Devlin and H. Johnsbråten, The Souslin Problem. VIII, 132 pages. 1974.

Vol. 406: Graphs and Combinatorics. Proceedings 1973. Edited by R. A. Bari and F. Harary. VIII, 355 pages. 1974.

Vol. 407: P. Berthelot, Cohomologie Cristalline des Schémas de Caracteristique p > o. II, 604 pages. 1974.

Vol. 408: J. Wermer, Potential Theory. VIII, 146 pages. 1974.

Vol. 409: Fonctions de Plusieurs Variables Complexes, Séminaire François Norguet 1970–1973. XIII, 612 pages. 1974.

Vol. 410: Séminaire Pierre Lelong (Analyse) Année 1972–1973. VI, 181 pages. 1974.

Vol. 411: Hypergraph Seminar. Ohio State University, 1972. Edited by C. Berge and D. Ray-Chaudhuri. IX, 287 pages. 1974.

Vol. 412: Classification of Algebraic Varieties and Compact Complex Manifolds. Proceedings 1974. Edited by H. Popp. V, 333 pages. 1974.

Vol. 413: M. Bruneau, Variation Totale d'une Fonction. XIV, 332 pages. 1974.

Vol. 414: T. Kambayashi, M. Miyanishi and M. Takeuchi, Unipotent Algebraic Groups. VI, 165 pages. 1974.

Vol. 415: Ordinary and Partial Differential Equations. Proceedings 1974. XVII, 447 pages. 1974.

Vol. 416: M. E. Taylor, Pseudo Differential Operators. IV, 155 pages. 1974.

Vol. 417: H. H. Keller, Differential Calculus in Locally Convex Spaces. XVI, 131 pages. 1974.

Vol. 418: Localization in Group Theory and Homotopy Theory and Related Topics. Battelle Seattle 1974 Seminar. Edited by P. J. Hilton. VI, 172 pages 1974.

Vol. 419: Topics in Analysis. Proceedings 1970. Edited by O. E. Lehto, I. S. Louhivaara, and R. H. Nevanlinna. XIII, 392 pages. 1974.

Vol. 420: Category Seminar. Proceedings 1972/73. Edited by G. M. Kelly. VI, 375 pages. 1974.

Vol. 421: V. Poénaru, Groupes Discrets. VI, 216 pages. 1974.

Vol. 422: J.-M. Lemaire, Algèbres Connexes et Homologie des Espaces de Lacets. XIV, 133 pages. 1974.

Vol. 423: S. S. Abhyankar and A. M. Sathaye, Geometric Theory of Algebraic Space Curves. XIV, 302 pages. 1974.

Vol. 424: L. Weiss and J. Wolfowitz, Maximum Probability Estimators and Related Topics. V, 106 pages. 1974.

Vol. 425: P. R. Chernoff and J. E. Marsden, Properties of Infinite Dimensional Hamiltonian Systems. IV, 160 pages. 1974.

Vol. 426: M. L. Silverstein, Symmetric Markov Processes. X, 287 pages. 1974.

Vol. 427: H. Omori, Infinite Dimensional Lie Transformation Groups. XII, 149 pages. 1974.

Vol. 428: Algebraic and Geometrical Methods in Topology, Proceedings 1973. Edited by L. F. McAuley. XI, 280 pages. 1974.

Vol. 429: L. Cohn, Analytic Theory of the Harish-Chandra C-Function. III, 154 pages. 1974.

Vol. 430: Constructive and Computational Methods for Differential and Integral Equations. Proceedings 1974. Edited by D. L. Colton and R. P. Gilbert. VII, 476 pages. 1974.

Vol. 431: Séminaire Bourbaki – vol. 1973/74. Exposés 436–452. IV, 347 pages. 1975.

Vol. 432: R. P. Pflug, Holomorphiegebiete, pseudokonvexe Gebiete und das Levi-Problem. VI, 210 Seiten. 1975.

Vol. 433: W. G. Faris, Self-Adjoint Operators. VII, 115 pages. 1975.

Vol. 434: P. Brenner, V. Thomée, and L. B. Wahlbin, Besov Spaces and Applications to Difference Methods for Initial Value Problems. II, 154 pages. 1975.

Vol. 435: C. F. Dunkl and D. E. Ramirez, Representations of Commutative Semitopological Semigroups. VI, 181 pages. 1975.

Vol. 436: L. Auslander and R. Tolimieri, Abelian Harmonic Analysis, Theta Functions and Function Algebras on a Nilmanifold. V, 99 pages. 1975.

Vol. 437: D. W. Masser, Elliptic Functions and Transcendence. XIV, 143 pages. 1975.

Vol. 438: Geometric Topology. Proceedings 1974. Edited by L. C. Glaser and T. B. Rushing. X, 459 pages. 1975.

Vol. 439: K. Ueno, Classification Theory of Algebraic Varieties and Compact Complex Spaces. XIX, 278 pages. 1975.

Vol. 440: R. K. Getoor, Markov Processes: Ray Processes and Right Processes. V, 118 pages. 1975.

Vol. 441: N. Jacobson, PI-Algebras. An Introduction. V, 115 pages. 1975.

Vol. 442: C. H. Wilcox, Scattering Theory for the d'Alembert Equation in Exterior Domains. III, 184 pages. 1975.

Vol. 443: M. Lazard, Commutative Formal Groups. II, 236 pages. 1975.

Vol. 444: F. van Oystaeyen, Prime Spectra in Non-Commutative Algebra. V, 128 pages. 1975.

Vol. 445: Model Theory and Topoi. Edited by F. W. Lawvere, C. Maurer, and G. C. Wraith. III, 354 pages. 1975.

Vol. 446: Partial Differential Equations and Related Topics. Proceedings 1974. Edited by J. A. Goldstein. IV, 389 pages. 1975.

Vol. 447: S. Toledo, Tableau Systems for First Order Number Theory and Certain Higher Order Theories. III, 339 pages. 1975.

Vol. 448: Spectral Theory and Differential Equations. Proceedings 1974. Edited by W. N. Everitt. XII, 321 pages. 1975.

Vol. 449: Hyperfunctions and Theoretical Physics. Proceedings 1973. Edited by F. Pham. IV, 218 pages. 1975.

Vol. 450: Algebra and Logic. Proceedings 1974. Edited by J. N. Crossley. VIII, 307 pages. 1975.

Vol. 451: Probabilistic Methods in Differential Equations. Proceedings 1974. Edited by M. A. Pinsky. VII, 162 pages. 1975.

Vol. 452: Combinatorial Mathematics III. Proceedings 1974. Edited by Anne Penfold Street and W. D. Wallis. IX, 233 pages. 1975.

Vol. 453: Logic Colloquium. Symposium on Logic Held at Boston, 1972–73. Edited by R. Parikh. IV, 251 pages. 1975.

Vol. 454: J. Hirschfeld and W. H. Wheeler, Forcing, Arithmetic, Division Rings. VII, 266 pages. 1975.

Vol. 455: H. Kraft, Kommutative algebraische Gruppen und Ringe. III, 163 Seiten. 1975.

Vol. 456: R. M. Fossum, P. A. Griffith, and I. Reiten, Trivial Extensions of Abelian Categories. Homological Algebra of Trivial Extensions of Abelian Categories with Applications to Ring Theory. XI, 122 pages. 1975.